Chemistry of Electronic Materials

Chemistry of Electronic Materials

Andrew R. Barron

MiDAS Green Innovations
2021

Cover image © 2020 by HQuality Video

First Printing: 2021

ISBN 978-1-8384167-1-3

MiDAS Green Innovation, Ltd
Swansea, SA1 8RD, UK

www.midasgreeninnovation.com

Dedication

To my wife, Merrie, for all your support, friendship and love especially during a wonderful year in "lockdown".

Table of Contents

Acknowledgements

I would like to thank colleagues and students who provided input to this book. In particular, the late Prof. William L. "Bill" Wilson of the Department of Electrical Engineering for his guidance during my first few years as a faculty member at Rice University and for inspiring me to disseminate my lectures as a textbook. The majority of this book was created from the postgraduate level course *CHEM 596:Chemistry of Electronic Materials* that I taught between 1997 and 1999. In particular, I would like to acknowledge students (all of whom have gone on to create careers of their own) who brought structure to my lectures and notes. These include: Zane Ball, Julie Francis, Andrea Keys, Carissa Smith, Scott Stokes, Angela Cindy Wei, Sarah Westcott, and Elizabeth Whitsitt.

I wish to acknowledge Prof. Thamraa Alshahrani for all her support with the *Energy Innovation* MSc I have the privilege to be co-developing in partnership with Princess Nourah Bint Abdulrahman University (PNU) in Saudi Arabia.

Last, but not least, I would like to thank my wife, Merrie, for putting up with me during the COVID-19 pandemic, in particular my absence due to hotel quarantine that extended what was originally planned as a short business trip to Qatar.

Preface

The intention of this text is not to provide a comprehensive reference to all aspects of semiconductor device fabrication. After all there are myriad books already on that subject by authors more expert than I. Nor is it intended to be a review of 'research results' that, irrespective of their promise, have not been adopted into mainstream production. In this case there are enough academics touting their own results with promises of 'saving the planet'. Instead it is aimed to provide a useful reference for those interested in the chemical aspects of the electronics industry.

In the early 1990s I found that I was becoming involved in research areas that were outside of my comfort area as an inorganic/organometallic chemist. Unlike many academics that joined the 'materials chemistry' bandwagon at that time, I wanted to ensure my research was relevant to the problems it was purporting to solve. My knowledge of semiconductor-based devices and processing involved in their manufacture had been limited to school and university physics courses, but even I knew that just making a compound containing gallium and arsenic didn't make it a precursor to GaAs if the end material contained significant carbon impurities. The text and reference books in the area are predominantly written for the electrical engineer, not so helpful to a chemist.

I, therefore, set out to bring together the type of information that a non-specialist needs to understand the field and be able to assess the relevance of research. This became critical in understanding the potential benefits of research that became the basis of my founding two start-up companies: Gallia, Inc. (passivation of GaAs devices) and Natcore Technology, Inc. (low cost solar cell manufacturing processes).

Once collected I found that others were equally in need of a suitable primer, and so the collected information evolved into a lecture course, *CHEM 596:Chemistry of Electronic Materials*, that I taught to chemistry undergraduate and graduate students at Rice University between 1997 and 1999. This became the basis of the 1st Edition of this book, published through now defunct OpenStax CNX (formally Connexions).

I have returned to update and focus the text because I am proud to be developing a new collaborative MSc program at Princess Nourah Bint

Abdulrahman University (PNU) in Saudi Arabia: MSc in *Energy Innovation*. As PNU is the largest women's university in the world the goal of the course is to encourage women into science and engineering areas that are the future in green energy and reducing carbon emissions. One critical part of this future will be solar energy that relies on photovoltaic device fabrication, and so an understanding of topics such as device design, wafer fabrication, thin film growth and metallization is important.

Finally, there is always confusion in the appropriate choice of nomenclature when referring to the Groups in which semiconductor elements are located within the Periodic Table. The IUPAC nomenclature would place silicon in Group 14, while, the more traditional, Group IV descriptor is used by the engineering community. In deference to the electronics field the older system will be used herein, albeit with the IUPAC nomenclature on parenthesis. Thus, silicon is Group IV (14) and gallium arsenide is a III-V (13-15) compound.

Abbreviations

ALD	Atomic layer deposition
ALE	Atomic layer epitaxy
CBD	Chemical bath deposition
CBE	Chemical beam epitaxy
CD	Critical dimension
CMOS	Complimentary metal-oxide-semiconductor
CSD	Chemical solution deposition
CVD	Chemical vapor deposition
CPU	Central processing unit
ED	Electroless deposition
FET	Field effect transistor
HBT	Heterojunction bipolar transistor
IC	Integrated circuit
LACVD	Laser-assisted chemical vapor deposition
LED	Light emitting diode
LEED	Low-energy electron diffraction
LPD	Liquid phase deposition
MBE	Molecular beam epitaxy
MESFET	Metal semiconductor field effect transistor
MISFET	Metal-insulator-semiconductor field effect transistor
MOCVD	Metal organic chemical vapor deposition
MOSFET	Metal-oxide-semiconductor field effect transistor
NA	Numerical aperture
PL	Photolithographic
PR	Photoresist
PVD	Physical vapor deposition
RHEED	Reflection high energy diffraction
SILAR	Successive ion layer adsorption and reaction
TSI	Top-surface-imaged
UHV	Ultra high vacuum

Chapter 1: Introduction to Semiconductors

A *semiconductor* material (e.g., silicon and gallium arsenide) has an electrical conductivity between that of a conductor (e.g., metallic copper) and an insulator (e.g., aluminum oxide). Unlike metals whose resistivity increases with increased temperature, semiconductors show an increased conductivity with increased temperature.

Semiconductors (at least those that are inorganic) are crystalline materials in which their atoms arranged in an ordered fashion. For example, silicon is a Group IV (14) element, which means it has four electrons in its outer or valence shell, resulting in a solid-state structure in which each atom has four covalent bonds, arranged in a tetrahedral formation about the atom center (Figure 1.1).

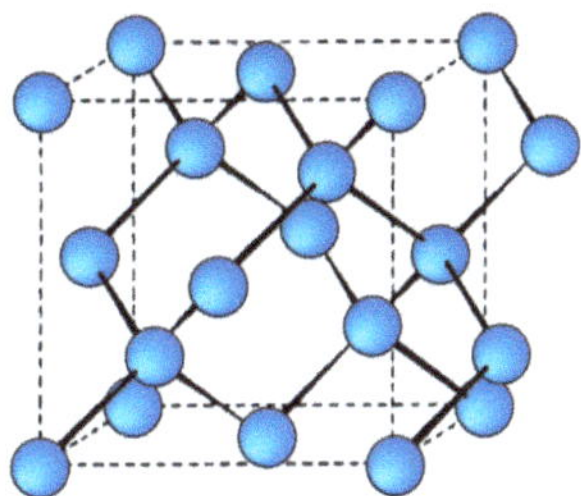

Figure 1.1: Unit cell structure of the diamond crystal lattice of elemental silicon showing the tetrahedral coordination of each atom.

Electronic structure of semiconductors

For the purposes of simplifying the following discussion of the electronic properties of semiconductors it is easiest to consider a 2-dimensional schematic representation of a piece of single-crystal silicon as being that shown in Figure 1.2. Each silicon atom shares its four valence electrons with valence electrons from four nearest neighbors, filling the shell to 8 electrons, and forming a stable, periodic structure. Once the atoms have been arranged in this manner, the outer valence electrons are no longer strongly bound to the individual host atom. Instead, the outer shells of all of the atoms blend together and form what is called a *band*. It is the movement of the electrons within this band that lead to electrical conductivity. This is not the complete story, however, for it turns out that due to quantum mechanical effects, there is not just one band which holds electrons, but several of them.

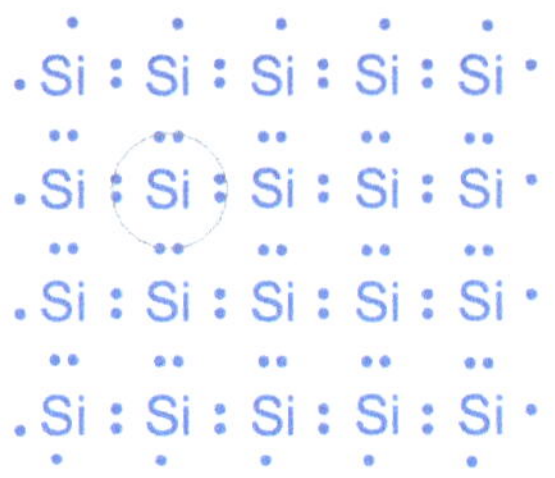

$$\begin{array}{ccccc}
\cdot\; \mathrm{Si} & \mathrm{Si} & \mathrm{Si} & \mathrm{Si} & \mathrm{Si}\; \cdot \\
\cdot\; \mathrm{Si} & \mathrm{Si} & \mathrm{Si} & \mathrm{Si} & \mathrm{Si}\; \cdot \\
\cdot\; \mathrm{Si} & \mathrm{Si} & \mathrm{Si} & \mathrm{Si} & \mathrm{Si}\; \cdot \\
\cdot\; \mathrm{Si} & \mathrm{Si} & \mathrm{Si} & \mathrm{Si} & \mathrm{Si}\; \cdot
\end{array}$$

Figure 1.2: A 2-D schematic representation of a silicon crystal. The circle represents the valence electron shell of one of the silicon atoms, which is comprised of 4 from the central atom and 4 more shared (one each) with the 4 neighboring silicon atoms.

Electrons are not only distributed throughout the solid crystal spatially, but they also have a distribution in energy as well. The potential energy function within the solid is periodic in nature. This potential function comes from the positively charged atomic nuclei, which are arranged in the crystal in a regular array. A detailed analysis of how electron wave functions, the mathematical abstraction, which one must use to describe how small quantum mechanical objects behave when they are in a periodic potential, gives rise to an energy distribution somewhat like that shown in Figure 1.3.

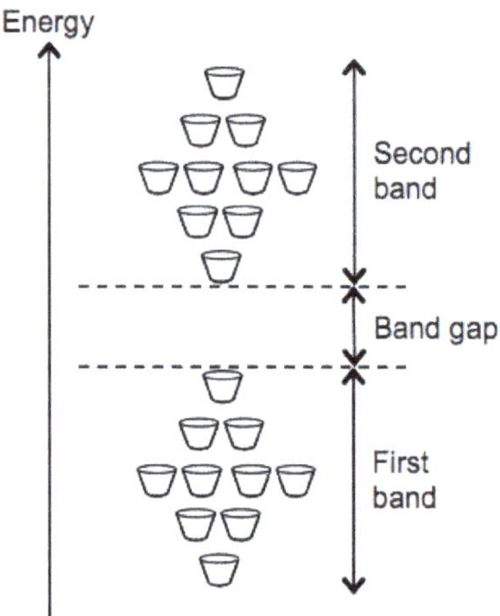

Figure 1.3. Schematic representation of the first two bands for a generic (periodic) solid, showing energy levels of the bands and the band gap, with the vertical axis representing electron energy. The cups represent particular possible energy level into which electrons may be placed.

Unlike the case for free electrons, in a periodic solid electrons are not free to take on any energy value they wish. They are forced into specific energy levels called *allowed states*, which are represented by the 'cups' in Figure 1.3. The allowed states are not distributed uniformly in energy; instead they are grouped into specific configurations called *energy bands*. There are no allowed levels at zero energy and for some distance above that. Moving up from zero energy, we then encounter the first energy band. At the bottom of the band there are very few *allowed states*, but as we move up in energy, the number of allowed states first increases (i.e., shown as an increase in the number of 'cups' at a particular energy in Figure 1.3), and then falls off again (i.e., shown as an decrease in the number of 'cups' at a particular energy in Figure 1.3). We then come to a region with no allowed states, called an energy *band gap*. Above the band gap, another band of allowed states exists. This goes on and on, with any given material having many such bands and band gaps.

It turns out that each band has exactly $2N$ allowed states in it, where N is the total number of atoms in the particular crystal sample. Thus, for Figure 1.3, since there are 10 cups in each band, it must represent the band structure of a crystal with just 5 atoms. Clearly, in a macroscopic piece of material there will be a very large number of bands. For example, in a 1 cm^3 piece of silicon there are 4.994×10^{22} atoms and hence 9.988×10^{22} allowed states.

Into these bands the *valence electrons* associated with the atoms are distribute, with the restriction that we can only put one electron into each allowed state (i.e., represented as a filled cup in Figure 1.3 in which one electron is placed in each cup). This is the result of something called the Pauli exclusion principle, which states, "no two electrons can have the same set of quantum numbers". If we make the logical assumption that the electrons will fill in the levels with the lowest energy first, and only go into higher lying levels if the ones below are already filled. This situation is shown in Figure 1.4, in which we have represented electrons as filled cups.

Figure 1.4 is an energy based one, showing how the electrons are distributed in energy, not how they are arranged spatially. Using this diagram we cannot show how the electrons would move if an electric field is applied. An external electric field will exert a force on the electrons and attempt to accelerate them. If the electrons are accelerated, then they must increase their kinetic energy. Unfortunately, there are no empty allowed states in the filled bands. An electron would have to jump all the way up into the next (empty) band in order to take on more energy. In silicon, the gap between the top of the highest most occupied band and the lowest unoccupied band (c.f., Figure 1.4) is 1.1 eV. Since 1 eV is defined as the potential energy gained by an electron moving across an electrical potential of one volt, and the distance over which an electron would normally move before it suffers a collision (mean free path) is only a few hundred angstroms ($Å = 1 \times 10^{-10}$ m), then a very large electric field, several hundred thousand V/cm, would be needed in order for the electron to pick up enough energy to 'jump the gap'. This makes it appear that silicon would be a very bad conductor of electricity, and indeed very pure silicon is very poor electrical conductor.

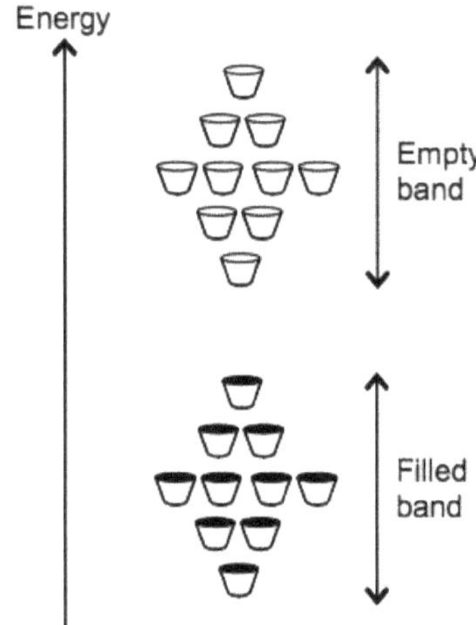

Figure 1.4: Schematic representation of the band structure of silicon, with the lower band full and the next empty.

A metal is an element with an upper band that is just half full of electrons. This is illustrated in Figure 1.5. Here we see that one band is full, and the next is half full. This would be the situation for the Group III (13) element aluminum for instance. If we apply an electric field to these carriers, those near the top of the distribution can indeed move into higher energy levels by acquiring some kinetic energy of motion, and easily move from one place to the next.

In comparison to a conductor, if there are no places for electrons to 'move' into, then how does silicon work as a *semiconductor*? Firstly, not all of the electrons are in the bottom bands. In silicon, unlike quartz (9.47 eV) or diamond (5.47 eV), the band gap between the top-most full band, the next empty one is not so large (1.1 eV), and so long as the silicon is not at absolute zero (0 K), some electrons near the top of the full band can acquire enough thermal energy that they can 'hop' the gap, and end up in the upper band, called the conduction band. This situation is shown in Figure 1.6.

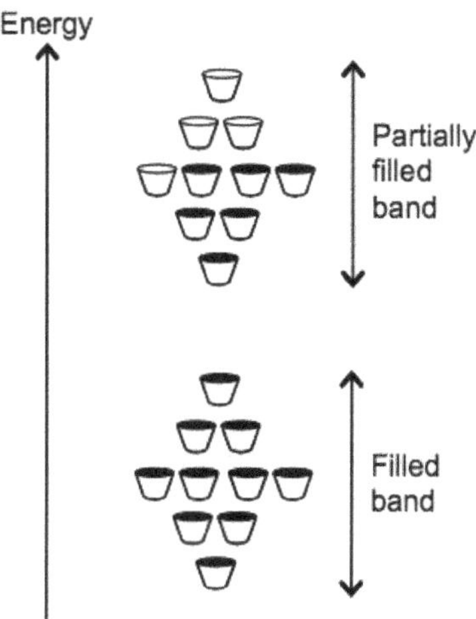

Figure 1.5: Schematic representation of the electron distribution for a metal or good conductor.

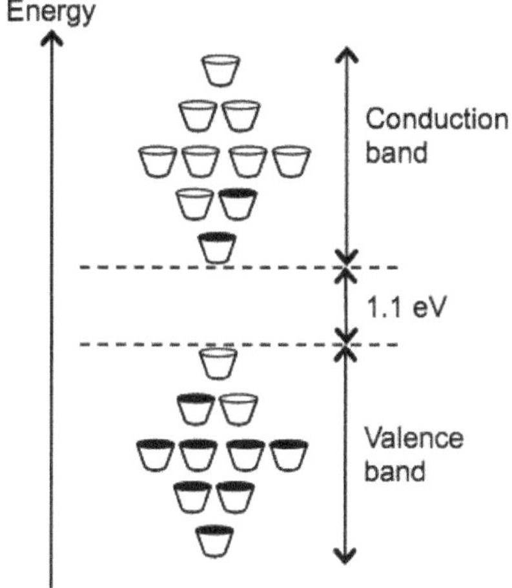

Figure 1.6: Schematic representation of the thermal excitation of electrons across the band gap in a semiconductor such as silicon.

In silicon at room temperature, roughly 10^{10} electrons per cubic centimeter are thermally excited across the band-gap at any one time, which is equivalent to approximately 1 electron per 5×10^{12} atoms. It should be noted that the excitation process is a continuous one. Electrons are being excited across the band, but then they fall back down into empty spots in the lower band, but on average 10^{10} electrons/cm^3 are thermally promoted at any given instant. But, is 10^{10} electrons/cm^3 enough to make silicon a conductor?

The mobility of electrons in silicon is about 1000 cm^2/V.s and the charge on an electron is 1.6×10^{-19} coulombs. Thus from

$$\sigma = nq\mu$$
$$= 10^{10}\,(1.6 \times 10^{-19})\,1000$$
$$= 1.6 \times 10^{-6}\ \text{mhos/cm}$$

a sample of silicon 1 cm long by (1 mm $\times$ 1 mm) square, it would have a resistance,

$$R = L/\sigma A$$
$$= 1/(1.6 \times 10^{-6})(0.1)^2$$
$$= 1.6 \times 10^{-6}\ \text{M}\Omega$$

which does not make it much of a "conductor".

Doped semiconductors

To see how we can make silicon a useful electronic material, we will have to go back to its crystal structure but substitute a few atoms of phosphorus for some of the silicon atoms (i.e., Figure 1.7). Phosphorus is a Group V (15) element, as compared with silicon, which is a Group IV (14) element, and as such the phosphorus atom has five outer or valence electrons, instead of the four in silicon. In a lattice composed mainly of silicon, the extra electron associated with the phosphorus atom has no 'mating' electron with which it can complete a shell, and so is left loosely dangling to the phosphorus atom, with relatively low binding energy. In fact, with the addition of just a little thermal energy (from the natural or latent heat of the crystal lattice) this electron can break free and be left to wander around the silicon atom freely.

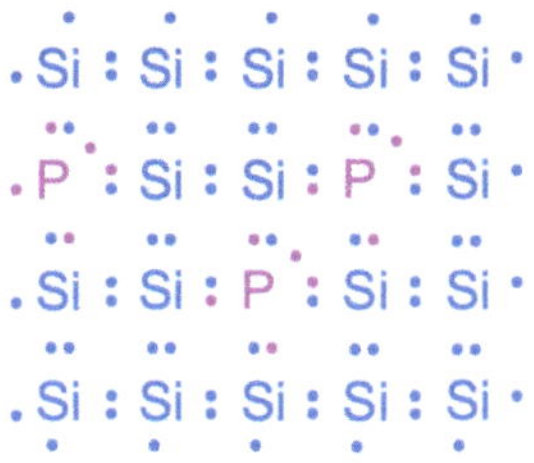

Figure 1.7: A 2-D schematic representation of a silicon crystal lattice *doped* with phosphorus in which 1 extra electron is present per phosphorus atom.

The 'energy band' picture can now be represented as Figure 1.8, in which the phosphorus atoms are represented by the purple cups that have new allowed energy levels which are formed within the silicon *band gap* near the bottom of the first empty band. These added energy levels are located close enough to the empty (or *conduction*) band, so that the electrons which they contain are easily excited up into the conduction band. There, they are free to move about and contribute to the electrical conductivity of the sample.

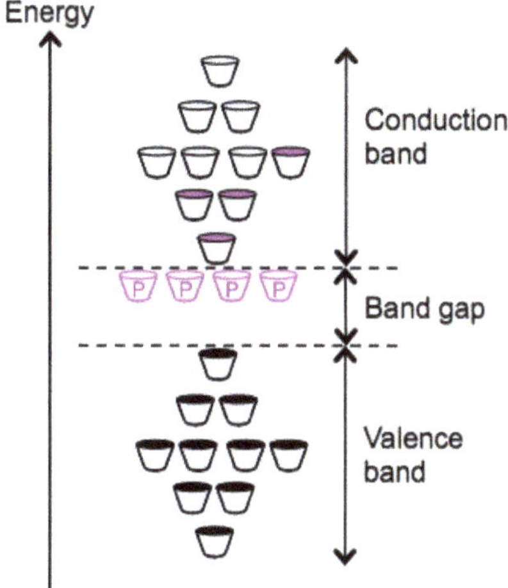

Figure 1.6: Schematic representation of silicon doped with phosphorus.

Since the electron has left the vicinity of the phosphorus atom to which it was originally associated, there is now one more proton than there are electrons at the atom, and hence it has a net positive charge of $1q$. We have represented this by creating an empty P-cup. Note that this positive charge is fixed at the site of the phosphorous atom called a *donor* since it 'donates' an electron up into the conduction band, and is not free to move about in the crystal.

How many phosphorus atoms are needed to significantly change the resistance of silicon? Suppose we wanted our 1 mm by 1 mm square sample to have a resistance of one ohm as opposed to more than 60 MΩ. Turning the resistance equation around we get,

$$\sigma = L/RA$$
$$= 1\ \Omega/1 \times (0.1)^2$$
$$= 100\ \text{mho/cm}$$

and hence, if we continue to assume an electron mobility of 1000 cm^2/volt.sec,

$$n = \sigma/q\mu$$
$$= 100/(1.6 \times 10^{-19})1000$$
$$= 6.25 \times 10^{17}\ \text{cm}^3$$

Adding 6.25×10^{17} phosphorus atoms per cubic centimeter might seem like a lot of phosphorus, until you realize that there are almost 10^{24} silicon atoms in a cubic centimeter and hence only one in every 79 thousand silicon atoms has to be changed into a phosphorus one to reduce the resistance of the sample from several MΩ down to only one Ω. It is this dramatic alteration of conductivity through *doping* that is the real power of semiconductors. It is also one of the great challenges of the semiconductor manufacturing industry, for it is necessary to maintain fantastic levels of control of the impurities in the material in order to predict and control their electrical properties. However, the addition of elements to the right of silicon (those with >4 valence electrons) is not the only method for doping.

Consider Figure 1.6, where there are a few 'empty' cups in the lower, almost full *valence* band. We will take another view of this band, from a somewhat different perspective. The following is a simplistic description of the effects of these empty energy levels. For a full understanding, a quantum mechanical description is required; however, what follows is intended to provide a rationalization of the results.

Consider Figure 1.7, which shows all of the electrons in the valence, or almost full band, and for simplicity show one missing electron. Applying an electric field, as shown by the arrow in Figure 1.7a. The field will try to move the (negatively charged) electrons to the left, but since the band is almost completely full, the only one that can move is the one right next to the empty spot, or *hole* as it is called. The result is movement of the hole to the right (Figure 1.7b), and subsequently, a short time after the one right next to the new *hole* moves (Figure 1.7c).

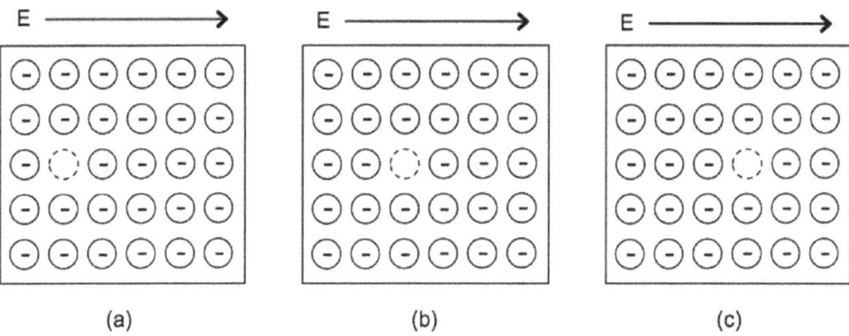

Figure 1.7: Idealized band full of electrons (a) with one missing and (b and c) the motion of the missing electron due to an applied electric field.

We can interpret this motion in two ways. One is that we have a net flow of negative charge to the left, or if we consider the effect of the aggregate of all the electrons in the band we could picture what is going on as a single positive charge, moving to the right (Figure 1.8).

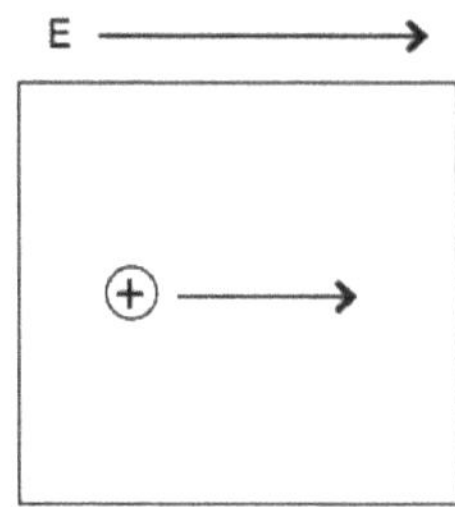

Figure 1.8: Motion of a hole due to an applied electric field.

Note that in either view we have the same net effect in the way the total net charge is transported through the sample. In the mostly negative charge picture (Figure 1.7), a net flow of negative charge to the left occurs, while in the single positive charge picture (Figure 1.8) a net flow of positive charge to the right occurs. Both give the same sign for the current.

The consequences of the empty spaces moving through the coordinated motion of electrons in an almost full band can be considered as being the motion of positive charges, moving wherever these empty spaces happen to be. These charge carriers (*holes*) can add to the total conduction of electricity in a semiconductor. Using ρ to represent the density (in cm^{-3} of spaces in the valence band and μ_e and μ_h to represent the mobility of electrons and holes respectively (they are usually not the same) we can modify to give the conductivity σ, when both electrons' holes are present,

$$\sigma = nq\mu_e + \rho q\mu_h$$

How can we get a sample of semiconductor with a lot of *holes* in it? Instead of phosphorus, the dopant element is chosen to the left of silicon in the Periodic Table, e.g., a group III (13) element such as boron as shown in Figure 1.9.

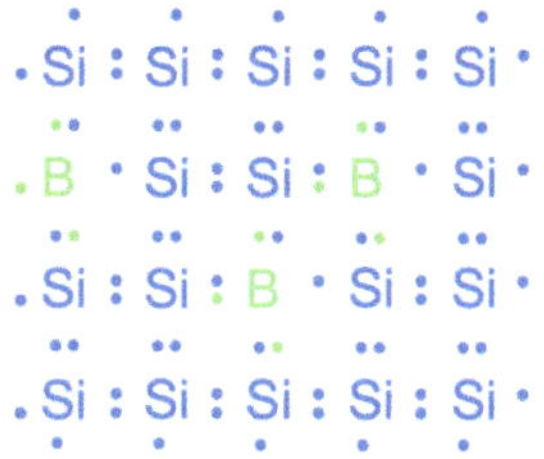

Figure 1.9: A 2-D schematic representation of a silicon crystal lattice *doped* with boron in which 1 less electron is present per boron atom.

In the energy diagram of the B-doped silicon (Figure 1.10) a set of new levels, introduced by the boron atoms, are located within the band gap, just a little way above the top of the almost full, or valence band. Electrons in the valence band can be thermally excited up into these new allowed levels, creating empty states, or holes, in the valence band. The excited electrons are stuck at the boron atom sites called *acceptors*, since they 'accept' an electron from the valence band, and hence act as fixed negative charges, localized there.

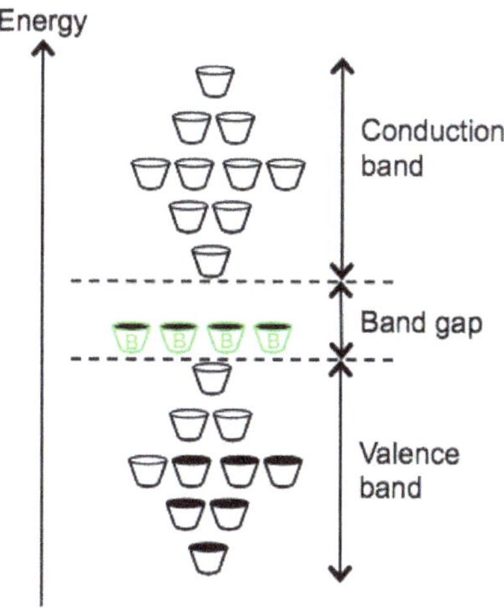

Figure 1.10: Schematic representation of silicon doped with boron.

A semiconductor, which is doped predominantly with acceptors, is called *p-type*, and most of the electrical conduction takes place through the motion of holes. A semiconductor, which is doped with donors, is called *n-type*, and conduction takes place mainly through the motion of electrons.

In n-type material, we can assume that all of the phosphorous atoms, or *donors*, are fully ionized when they are present in the silicon structure. Since the number of donors is usually much greater than the native, or intrinsic electron concentration ($\approx 10^{10}$ cm^{-3}) if N_d is the density of donors in the material, then n, the electron concentration, $\approx N_d$. If an electron deficient material such as boron is present (*p-type* silicon) the *hole* concentration is just $\approx N_a$ the concentration of *acceptors*, since these atoms 'accept' electrons from the valence band.

If both donors and acceptors are in the material, then whichever one has the higher concentration dominates. This is called *compensation*. If there are more donors than acceptors then the material is *n-type* and $n \approx N_d - N_a$. If there are more acceptors than donors then the material is *p-type* and $p \approx N_a - N_d$. It should be noted that in most compensated material, one type of impurity usually has a much greater (several order of magnitude) concentration than the other, and so the subtraction process described above usually does not change things very much, e.g., $10^{18} - 10^{16} \approx 10^{18}$.

The product of the electron and hole concentration in a material must remain a constant. In silicon at room temperature:

$$np \equiv n_i^2 \approx 10^{20} \text{ cm}^{-3}$$

Thus, if we have an n-type sample of silicon doped with 10^{17} donors/cm^3, then n, the electron concentration is just p, the hole concentration, is $10^{20}/10^{17} = 10^3$/cm^3. The carriers that dominate a material are called *majority carriers*, which would be the electrons in the above example. The other carriers are called *minority carriers* (the holes in this example) and while 10^3 might not seem like much compared to 10^{17} the presence of minority carriers is still quite important and cannot be ignored. Note that if the material is undoped, then it must be that $n = p$ and $n = p = 10^{10}$.

The picture of 'cups' of different allowed energy levels is useful for gaining a pictorial understanding a semiconductor, but becomes somewhat awkward when you want to start looking at devices which are made up of both n and p type silicon. Thus, another way of describing is called a band diagram (Figure 1.11).

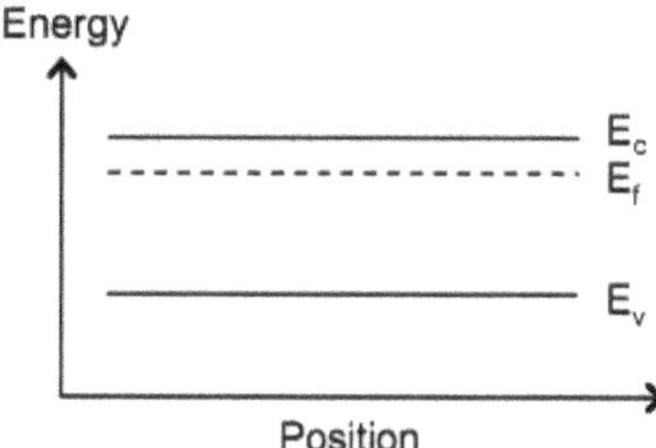

Figure 1.11: An electron band-diagram for a semiconductor.

A band diagram is just a representation of the energy as a function of position with a semiconductor device. In a band diagram, positive energy for electrons is upward, while for holes, positive energy is downwards. That is, if an electron moves upward, its potential energy increases just as with a normal mass in a gravitational field. Also, just as a mass will 'fall down' if given a chance, an electron will move down a slope shown in a band diagram. On the other hand, holes gain energy by moving downward and so they have a tendency to 'float' upward if given the chance; much like a bubble in a liquid. The line labeled E_c in Figure 1.11 shows the edge of the conduction band, or the bottom of the lowest unoccupied allowed band, while E_v is the top edge of the valence band, or highest occupied band. The band gap, E_g, for the material is $E_c - E_v$. The dotted line labeled E_f is called the *Fermi level* and it is indicative of the chemical equilibrium energy of the material, and also the type and number of carriers in the material. Note that there is no zero energy level on a band diagram. Either the Fermi level or one or other of the band edges is used as a reference level in lieu of knowing exactly where 'zero energy' is located.

The distance (in energy) between the Fermi level and either E_c and E_v gives us information concerning the density of electrons (n) and holes (p) in that region of the semiconductor material:

$$n = N_c e^{-\left(\frac{E_c - E_f}{kT}\right)}$$

$$p = N_v e^{-\left(\frac{E_f - E_v}{kT}\right)}$$

Both N_c and N_v are constants that depend on the material, but are typically on the order of $10^{19}/cm^3$. The expression in the denominator of the exponential is the Boltzman's constant ($k = 8.63 \times 10^{-5}$ eV/K) multiplied by the temperature T (in Kelvin) of the material. At room temperature $kT = \frac{1}{40}$ eV. It should be noted that the bigger the number in the exponent, the fewer carriers are in the material. Thus, based upon the top expression it is important for E_f must not be too far away from the conduction band in a n-type material, while in a p-type material the Fermi level, E_f must be down close to the valence band. The closer E_f gets to E_c the more electrons and the closer E_f gets to E_v, the more holes. Figure 1.12 therefore must be for a sample of n-type material. Note also that if a sample is heavily doped (i.e., N_d is known) and from $n \approx N_d$ the distance of the Fermi level from the conduction band is calculated,

$$E_f - E_f = kT \ln\left(\frac{N_c}{N_d}\right)$$

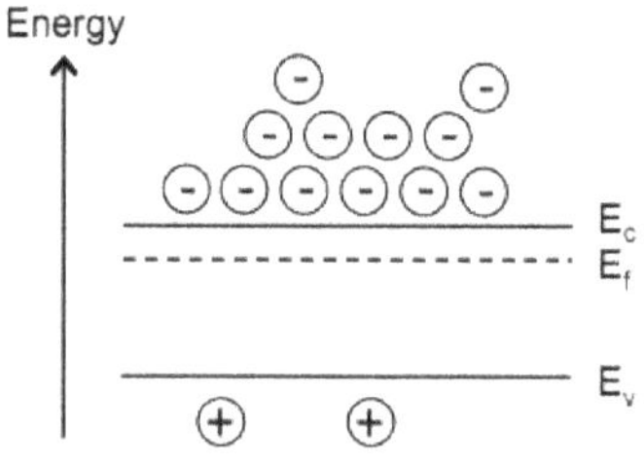

Figure 1.12: Band diagram for an n-type semiconductor.

To help further in our ability to picture what is going on in Figure 1.12, it is common to add small signed circles to indicate the presence of mobile electrons and holes in the material. Note that the electrons are spread out in energy. From our 'cup' picture we know they like to stay in the lower energy states if possible (Figure 1.4), but some will be distributed into the higher levels as well (Figure 1.6). What is distorted in Figure 1.12 is the scale. The band-gap for silicon is 1.1 eV, while the actual spread of the electrons would probably only be a few tenths of an eV, not nearly as much as is shown in Figure 1.12.

It should be noted that for both n and p-type material there are also a few "minority" carriers, or carriers of the opposite type, which arise from thermal generation across the band-gap. These are represented by the few positive charged holes in the valence band in Figure 1.12.

Diffusion

When electrons are injected across a junction, they move away from the junction region by a diffusion process, while at the same time, some of them are disappearing because they are minority carriers (electrons in basically p-type material) and so there are lots of holes around for them to recombine with. This is all shown schematically in Figure 1.13.

Diffusion may be quantified to create an expression for the electron distribution within the p-region. Imagine that a series of bins, each with a different number of electrons in them. To simplify the model only three bins will be considered. Over time, all of the electrons would flow out of their bins into the neighboring ones. Since there is no reason to expect the electrons to favor one side over the other, it can be assumed that exactly half leave by each side. This is all shown in Figure 1.14. Imagine there are 4, 6, and 8 electrons respectively in each of the bins. After the required 'emptying time', a net flux of exactly one electron across each boundary as shown, i.e., 3 − 2 or 4-3, etc.

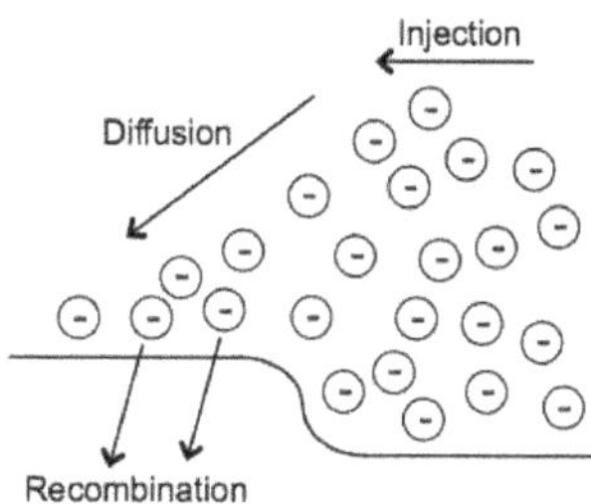

Figure 1.13: Processes involved in electron transport across a p-n junction.

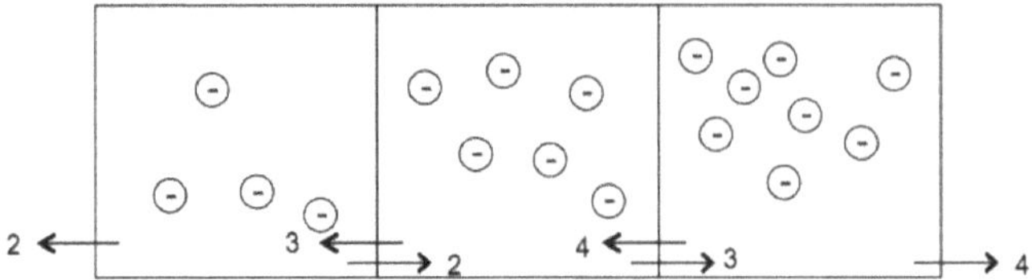

Figure 1.14: A schematic representation of a diffusion problem.

Now let's raise the number of electrons to 8, 12 and 16 respectively (Figure 1.15). We find that the net flux across each boundary is now 2 electrons per emptying time, rather than one. Note that the gradient (slope) of the concentration in the boxes has also doubled from one per box to two per box. This leads to the statement that the *flux* of carriers is proportional to the *gradient* of their density. This is stated formally in what is known as Fick's First Law of Diffusion,

$$\text{Flux} = (-D_e)\,\frac{d\,n\,(x)}{d\,x}$$

where, D_e is the diffusion coefficient, and n is the number of electrons at position x. Since we are talking about the motion of electrons, this diffusion flux must give rise to a current density $J_{e\text{diff}}$. Since an electron has a charge (q) associated with it,

$$J_{e_{\text{diff}}} = qD_e\,\frac{d\,n}{d\,x}$$

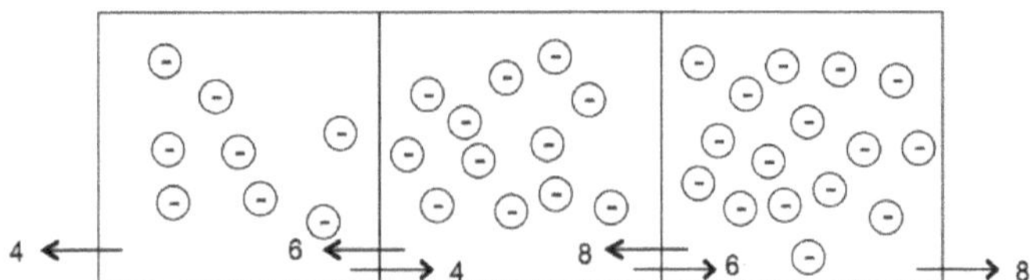

Figure 1.15: A schematic representation of electron diffusion from bins.

For a volume (V), which is filled with some charge (Q), the sum of all of the current density flowing out of the volume must be equal to the time rate of decrease of the charge within that volume:

$$\oint_S J \, \mathrm{d}S = -\frac{\mathrm{d}Q}{\mathrm{d}t}$$

Q can be written as,

$$Q = \oint_V \rho(v) \, \mathrm{d}V$$

which is a volume integral of the charge density (ρ) over the volume (V), using Gauss' theorem,

$$\oint_S J \, \mathrm{d}S = \int \mathrm{div}\,(J) \, \mathrm{d}V$$

So, combining the above we have,

$$\int \mathrm{div}\,(J) \, \mathrm{d}V = -\int \frac{\mathrm{d}\rho}{\mathrm{d}t} \, \mathrm{d}V$$

Shrinking the volume V down to a point, which means the quantities inside the integral must be equal, and we have the differential form of the continuity equation (in one dimension),

$$\mathrm{div}\,(J) = \frac{\partial J}{\partial x}$$
$$= -\frac{\mathrm{d}\,\rho(x)}{\mathrm{d}t}$$

What about the electrons?

The electrons that have been injected across the junction in Figure 1.13 are called *excess minority carriers*, because they are electrons in a p-region (hence minority) but their concentration is greater than what they would be if they were in a sample of p-type material at equilibrium. Their concentration is designated as $n'(x,t)$ since they could change with both time (t) and position (x). There are two ways in which $n'(x,t)$ can change with time. One would be if we were to stop injecting electrons in from the n-side of the junction. A reasonable way to account for the decay, which would occur if we were not supplying electrons, would be:

$$\frac{d}{dt} n'(x,t) = -\frac{n'(x,t)}{\tau_r}$$

where τ_r called the minority carrier *recombination lifetime*. If starting out with an excess minority carrier concentration n_0' at $t = 0$, then $n'(x,t)$ will be,

$$n'(x,t) = n'_0 e^{\frac{-t}{\tau_r}}$$

But, the electron concentration can also change because of electrons flowing into or out of the region x. The electron concentration $n'(x,t)$ is just $\rho(x,t)q$. Thus, due to electron flow,

$$\frac{d}{dt} n'(x,t) \;=\; \frac{1}{q} \frac{d\,\rho(x,t)}{dt}$$
$$=\; \frac{1}{q} \operatorname{div}\,(J(x,t))$$

Reducing the divergence to one dimension (we just have a $\partial J/\partial x$) gives,

$$\frac{d}{dt} n'(x,t) = D_e \frac{d^2 n'(x,t)}{dx^2}$$

Combining (electrons will, after all, suffer from both recombination and diffusion) gives:

$$\frac{d}{dt} n'(x,t) = D_e \frac{d^2 n'(x,t)}{dx^2} - \frac{n'(x,t)}{\tau_r}$$

This is a somewhat specialized form of the *ambipolar diffusion equation*. At steady state the time derivative on the LHS of the proceeding equation is zero, and so $n'(x,t)$ becomes $n'(x)$ since there is no time variation to worry about:

$$\frac{d^2}{d\,t^2}\,n'\,(x) - \frac{1}{D_e \tau_r}\,n'\,(x) = 0$$

At the boundary conditions that n'(0) = n_0 (the concentration of excess electrons just at the start of the diffusion region) and n'(x) $\rightarrow$ 0 as x $\rightarrow$ ∞ (the excess carriers becomes zero at a great distance from the junction) then:

$$n\,(x) = n_0 e^{-\frac{x}{\sqrt{D_e \tau_r}}}$$

The expression in the radical, $\sqrt{D_e \tau_r}$, is the electron diffusion length, L_e, and gives an idea as to how far away from the junction the excess electrons will exist before they have all recombined. The diffusion coefficient for electrons (D_e) in silicon is 25 cm^2/sec and the minority carrier lifetime is usually around a microsecond, which correlates to a very small distance, e.g.,

$$\begin{aligned}
L_e \;&=\; \sqrt{D_e \tau_r} \\
&=\; \sqrt{25 \times 10^{-6}} \\
&=\; 5 \times 10^{-3}\ \text{cm}
\end{aligned}$$

Chapter 2: Structures of Element and Compound Semiconductors

A single crystal of either an elemental (e.g., silicon) or compound (e.g., gallium arsenide) semiconductor forms the basis of almost all semiconductor devices. The ability to control the electronic and opto-electronic properties of these materials is based on an understanding of their structure. In addition, the metals and many of the insulators employed within a microelectronic device are also crystalline.

Group IV (14) elements

The diamond cubic structure consists of two interpenetrating *face-centered cubic* (*fcc*) lattices, with one off set $^1/_4$ of a cube along the cube diagonal. It may also be described as face centered cubic lattice in which half of the tetrahedral sites are filled while all the octahedral sites remain vacant. The diamond cubic unit cell is shown in Figure 2.1. Each of the atoms (e.g., C) is four-coordinate, and the shortest interatomic distance (i.e., C-C) may be determined from the unit cell parameter (*a*):

$$\text{C-C} \;=\; a\frac{\sqrt{3}}{4} \;\approx\; 0.422\,a$$

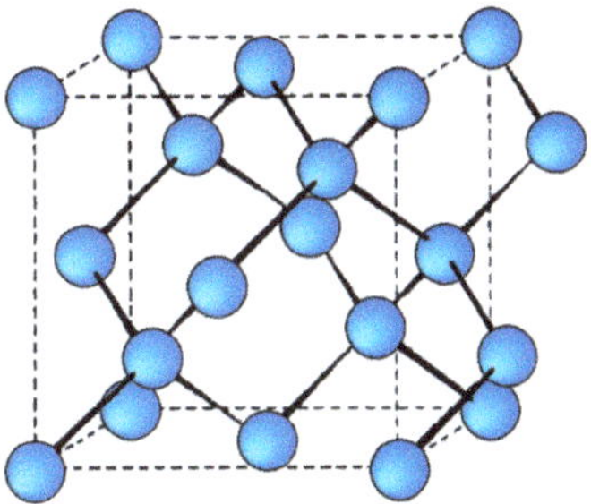

Figure 2.1: Unit cell (dashed lines) structure of a diamond cubic lattice showing the two-interpenetrating face-centered cubic lattices.

All the semiconducting phases of the group IV (14) elements, C (diamond), Si, Ge, and α-Sn, adopt the *diamond cubic* structure (Figure 2.1). Their lattice constants (*a*, Å) and densities (ρ, g/cm^3) are given in Table 2.1.

Element	Lattice parameter, a (Å)	Density (g/cm^3)
Carbon (diamond)	3.56683(1)	3.51525
Silicon	5.4310201(3)	2.319002
Germanium	5.657906(1)	5.3234
Tin (α-Sn)	6.4892(1)	7.285

Table 2.1: Lattice parameters and densities (measured at 298 K) for the diamond cubic forms of the Group IV (14) elements.

As would be expected based upon the atomic radii, the lattice parameter increase in the order C < Si < Ge < α-Sn. The similarity in lattice parameters for silicon and germanium (ca. 4%) means that they form a continuous series of *solid solutions* with gradually varying lattice parameters. It is worth noting the high degree of accuracy that the lattice parameters are known for high purity crystals of these elements. In addition, it is important to note the temperature at which structural measurements are made, since the lattice parameters are temperature dependent (Figure 2.2). The lattice constant (a), in Å, for high purity silicon may be calculated for any temperature (T) over the temperature range 293-1073 K by:

$$a_T = 5.4304 + 1.8138 \times 10^{-5} (T - 298.15 \text{ K}) + 1.542 \times 10^{-9} (T - 298.15 \text{ K})$$

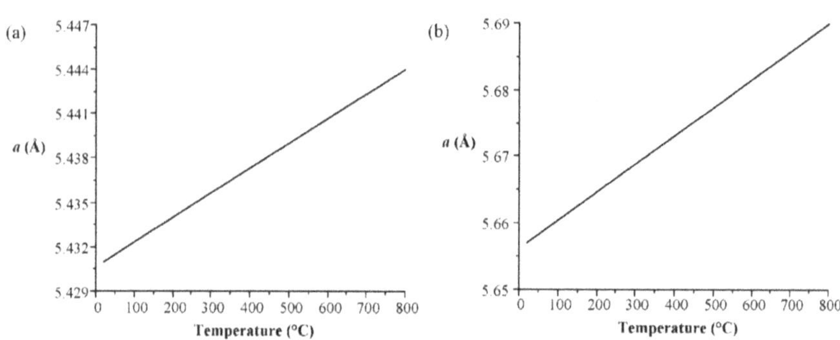

Figure 2.2: Temperature dependence of the lattice parameter for (a) silicon and (b) germanium.

Even though the diamond cubic forms of silicon and germanium are the only forms of direct interest to semiconductor devices, each exists in numerous crystalline high pressure and meta-stable forms. These are described along with their inter-conversion in Table 2.2.

Phase	Structure	Remarks
Si I	Diamond cubic	Stable at normal pressure
Si II	Grey tin structure	Formed from Si I or Si V >14 GPa
Si III	Cubic	Metastable, formed from Si II >10 GPa
Si IV	Hexagonal	
Si V	Unidentified	Stable >34 GPa, formed from Si II >16 GPa
Si VI	Hexagonal close packed	Stable above 45 GPa
Ge I	Diamond cubic	Low-pressure phase
Ge II	β-tin structure	Formed from Ge I above 10 GPa
Ge III	Tetragonal	Formed by quenching Ge II at low pressure
Ge IV	Body centered cubic	Formed by quenching Ge II to 1 atm at 200 K

Table 2.2: High pressure and metastable phases of silicon and germanium.

Group III-V (13-15) compounds

The stable phases for the arsenides, phosphides and antimonides of aluminum, gallium and indium all exhibit *zinc blende* structures (Figure 2.3), which is named after its archetype, a common mineral form of zinc sulfide (ZnS). As with the diamond lattice, zinc blende consists of the two interpenetrating *fcc* lattices; however, in zinc blende one lattice consists of one of the types of atoms (i.e., Ga in GaAs), and the other lattice is of the second type of atom (i.e., As in GaAs). It may also be described as face centered cubic lattice of As atoms in which half of the tetrahedral sites are filled with Ga atoms. All the atoms in a zinc blende structure are 4-coordinate. The zinc blende unit cell is shown in Figure 2.3.

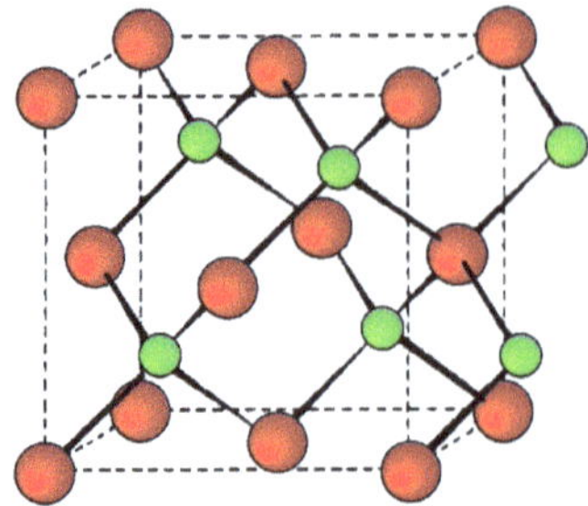

Figure 2.3: Unit cell structure of a zinc blende (ZnS) lattice. Zinc atoms are shown in green, sulfur atoms shown in red, and the dashed lines show the unit cell.

A number of inter-atomic distances may be calculated for any material with a zinc blende unit cell using the lattice parameter (a):

$$Zn\text{-}S = a\frac{\sqrt{3}}{4} \approx 0.422\,a$$

$$Zn\text{-}Zn = S\text{-}S = \frac{a}{\sqrt{2}} \approx 0.707\,a$$

In contrast, the nitrides (e.g., GaN) are found as wurtzite structures (Figure 2.4), which is a hexagonal form of the zinc sulfide. It is identical in the number of and types of atoms, but it is built from two interpenetrating *hcp* lattices as opposed to the *fcc* lattices in zinc blende. All the atoms in a wurtzite structure are 4-coordinate. The wurtzite unit cell is shown in Figure 2.4. A number of inter atomic distances may be calculated for any material with a wurtzite cell using the lattice parameter (a):

$$Zn\text{-}S = a\sqrt{3/8} = 0.612\,a = \frac{3c}{8} = 0.375\,c$$

$$Zn\text{-}Zn = S\text{-}S = a = 1.632\,c$$

It should be noted that these formulae do not necessarily apply when the ratio a/c is different from the ideal value of 1.632.

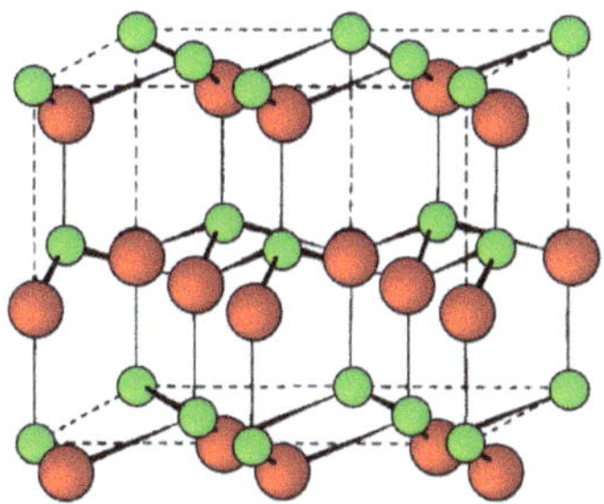

Figure 2.4: Unit cell structure of a wurtzite lattice. Zinc atoms are shown in green and sulfur atoms shown in red.

The structure, lattice parameters, and densities of the III-V compounds are given in Table 2.3. It is worth noting that contrary to expectation the lattice parameter of the gallium compounds is smaller than their aluminum homolog (i.e., GaAs a = 5.653 Å; AlAs a = 5.660 Å). As with the group IV elements the lattice parameters are highly temperature dependent; however, additional variation arises from any deviation from absolute stoichiometry. These effects are shown in Figure 2.5.

Compound	Structure	Lattice parameter (Å)	Density (g/cm³)
AlN	Wurtzite	a = 3.11(1), c = 4.98(1)	3.255
AlP	Zinc blende	a = 5.4635(4)	2.40(1)
AlAs	Zinc blende	a = 5.660	3.760
AlSb	Zinc blende	a = 6.1355(1)	4.26
GaN	Wurtzite	a = 3.190, c = 5.187	6.1
GaP	Zinc blende	a = 5.4505(2)	4.138
GaAs	Zinc blende	a = 5.65325(2)	5.3176(3)
InN	Wurtzite	a = 3.5446, c = 5.7034	6.81
InP	Zinc blende	a = 5.868(1)	4.81
InAs	Zinc blende	a = 6.0583	5.667
InSb	Zinc blende	a = 6.47937	5.7747(4)

Table 2.3: Lattice parameters and densities (measured at 298 K) for the III-V (13-15) compound semiconductors. Estimated standard deviations are given in parentheses.

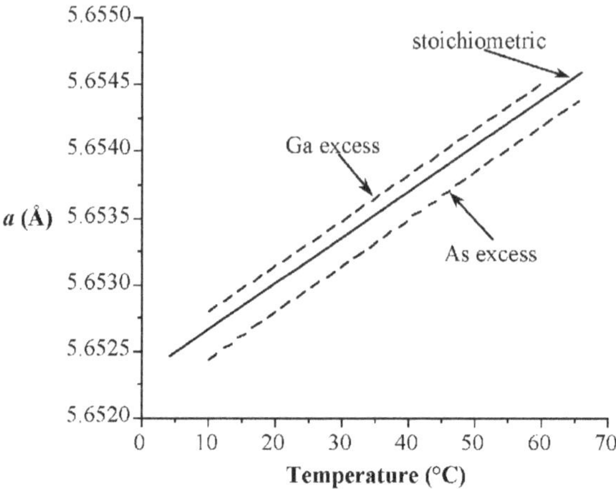

Figure 2.5: Temperature dependence of the lattice parameter for stoichiometric GaAs and crystals with either gallium or arsenic excess.

The homogeneity of structures allows for alloys of a wide range of solid solutions to be formed between III-V compounds in almost any combination. Two classes of ternary alloys are formed: III_x-III_{1-x}-V (e.g., Al_x-Ga_{1-x}-As) and III-V_{1-x}-V_x (e.g., Ga-As_{1-x}-P_x). While quaternary alloys of the type III_x-III_{1-x}-V_y-V_{1-y} allow for the growth of materials with similar lattice parameters, but a broad range of band gaps. A very important ternary alloy, especially in optoelectronic applications, is Al_x-Ga_{1-x}-As (often simply referred simply as AlGaAs) and its lattice parameter (a) is directly related to the composition (x).

$$a = 5.6533 + 0.0078 \text{ x}$$

Not all of the III-V compounds have well characterized high-pressure phases; however, in each case where a high-pressure phase is observed the coordination number of both the group III and group V element increases from four to six. Thus, AlP undergoes a zinc blende to rock salt transformation at high pressure above 170 kbar, while AlSb and GaAs form orthorhombic distorted rock salt structures above 77 and 172 kbar, respectively. An orthorhombic structure is proposed for the high-pressure form of InP (>133 kbar). Indium arsenide (InAs) undergoes two-phase transformations. The zinc blende structure is converted to a rock salt structure above 77 kbar, which in turn forms a β-tin structure above 170 kbar.

Group II-VI (12-16) compounds

The structures of the II-VI compound semiconductors are less predictable than those of the III-V compounds, and while zinc blende structure exists for almost all of the compounds there is a stronger tendency towards the hexagonal wurtzite form. In several cases the zinc blende structure is observed under ambient conditions, but may be converted to the wurtzite form upon heating. In general the wurtzite form predominates with the smaller anions (e.g., oxides), while the zinc blende becomes the more stable phase for the larger anions (e.g., tellurides). One exception is mercury sulfide (HgS) that is the archetype for the trigonal cinnabar phase. Table 2.4 lists the stable phase of the chalcogenides of zinc, cadmium and mercury, along with their high temperature phases where applicable. Solid solutions of the II-VI compounds are not as easily formed as for the III-V compounds; however, two important examples are ZnS_xSe_{1-x} and $Cd_xHg_{1-x}Te$.

The zinc chalcogenides all transform to a cesium chloride structure under high pressures, while the cadmium compounds all form *rock salt* high-pressure phases (Figure 2.6), which as its name implies the archetypal rock salt structure is NaCl (table salt). In common with the zinc blende structure, rock salt consists of two interpenetrating face-centered cubic lattices. However, the second lattice is off set $^1/_2a$ along the unit cell axis. It may also be described as face centered cubic lattice in which all of the octahedral sites are filled, while all the tetrahedral sites remain vacant, and thus each of the atoms in the rock salt structure are 6-coordinate. The rock salt unit cell is shown in Figure 2.6.

Compound	Structure	Lattice parameter (Å)	Density (g/cm³)
ZnS	Zinc blende	$a = 5.410$	4.075
	Wurtzite	$a = 3.822, c = 6.260$	4.087
ZnSe	Zinc blende	$a = 5.668$	5.27
ZnTe	Zinc blende	$a = 6.10$	5.636
CdS	Wurtzite	$a = 4.136, c = 6.714$	4.82
CdSe	Wurtzite	$a = 4.300, c = 7.011$	5.81
CdTe	Zinc blende	$a = 6.482$	5.87
HgS	Cinnabar	$a = 4.149, c = 9.495$	8.1
	Zinc blende	$a = 5.851$	7.73
HgSe	Zinc blende	$a = 6.085$	8.25
HgTe	Zinc blende	$a = 6.46$	8.07

Table 2.4: Lattice parameters and densities (measured at 298 K) for the II-VI (12-16) compound semiconductors.

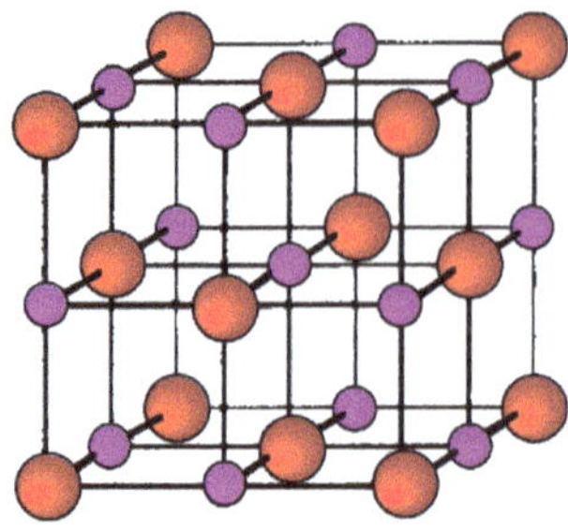

Figure 2.7: Unit cell structure of a rock salt lattice. Sodium ions are shown in purple and chloride ions are shown in red.

A number of inter-atomic distances may be calculated for any material with a rock salt structure using the lattice parameter (a):

$$Na\text{-}Cl \; = \; \frac{a}{2} \; \approx \; 0.5\,a$$

$$Na\text{-}Na \; = \; Cl\text{-}Cl \; = \; \frac{a}{\sqrt{2}} \; \approx \; 0.707\,a$$

I-III-VI$_2$ (11-13-16) compounds

Nearly all I-III-VI$_2$ compounds (e.g., $CuInS_2$) at room temperature adopt the chalcopyrite structure (Figure 2.7). The cell constants and densities are given in Table 2.5. Although there are few reports of high temperature or high-pressure phases, $AgInS_2$ has been shown to exist as a high temperature orthorhombic polymorph (a = 6.954, b = 8.264, and c = 6.683 Å), and $AgInTe_2$ forms a cubic phase at high pressures.

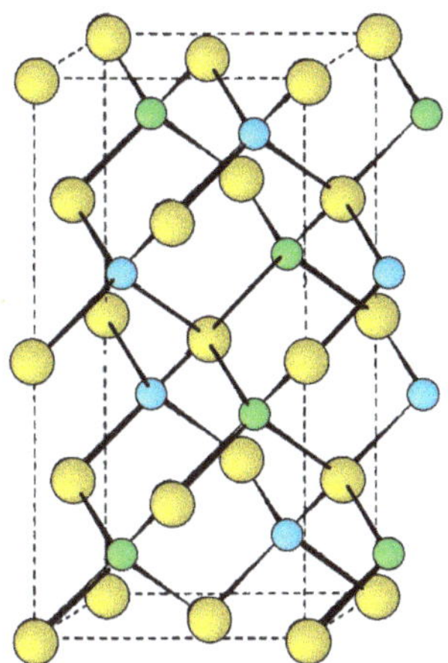

Figure 2.7: Unit cell structure of a chalcopyrite (CuFeS$_2$) lattice. Copper atoms are shown in blue; iron atoms are shown in green; and sulfur atoms are shown in yellow. The dashed lines show the unit cell.

Of the I-III-VI$_2$ compounds, the copper indium chalcogenides (CuInE$_2$) are certainly the most studied for their application in solar cells. One of the advantages of the copper indium chalcogenide compounds is the formation of solid solutions (alloys) of the formula CuInE$_{2-x}$E'$_x$, where the composition variable (x) varies from 0 to 2. The CuInS$_{2-x}$Se$_x$ and CuInSe$_{2-x}$Te$_x$ systems have also been examined, as has the CuGa$_y$In$_{1-y}$S$_{2-x}$Se$_x$ quaternary system. As would be expected from a consideration of the relative ionic radii of the chalcogenides the lattice parameters of the CuInS$_{2-x}$Se$_x$ alloy should increase with increased selenium content. Vergard's law requires the lattice constant for a linear solution of two semiconductors to vary linearly with composition (e.g., as is observed for Al$_x$Ga$_{1-x}$As), however, the variation of the tetragonal lattice constants (*a* and *c*) with composition for CuInS$_{2-x}$S$_x$ are best described by the parabolic relationships.

$$a = 5.532 + 0.0801 \, x + 0.0260 \, x^2$$

$$c = 11.156 + 0.1204 \, x + 0.0611 \, x^2$$

Compound	Lattice parameter a (Å)	Lattice parameter c (Å)	Density (g.cm^3)
CuAlS$_2$	5.32	10.430	3.45
CuAlSe$_2$	5.61	10.92	4.69

$CuAlTe_2$	5.96	11.77	5.47
$CuGaS_2$	5.35	10.46	4.38
$CuGaSe_2$	5.61	11.00	5.57
$CuGaTe_2$	6.00	11.93	5.95
$CuInS_2$	5.52	11.08	4.74
$CuInSe_2$	5.78	11.55	5.77
$CuInTe_2$	6.17	12.34	6.10
$AgAlS_2$	6.30	11.84	6.15
$AgGaS_2$	5.75	10.29	4.70
$AgGaSe_2$	5.98	10.88	5.70
$AgGaTe_2$	6.29	11.95	6.08
$AgInS_2$	5.82	11.17	4.97
$AgInSe_2$	6.095	11.69	5.82
$AgInTe_2$	6.43	12.59	6.96

Table 2.5: Chalcopyrite lattice parameters and densities (measured at 298 K) for the I-III-VI compound semiconductors. Lattice parameters are given for the tetragonal cell.

A similar relationship is observed for the $CuInSe_{2-x}Te_x$ alloys.

$$a = 5.783 + 0.1560\,x + 0.0212\,x^2$$

$$c = 11.628 + 0.3340\,x + 0.0277\,x^2$$

The large difference in ionic radii between S and Te (0.37 Å) prevents formation of solid solutions in the $CuInS_{2-x}Te_x$ system, however, the single alloy $CuInS_{1.5}Te_{0.5}$ has been reported.

Orientation effects

Once single crystals of high purity silicon or gallium arsenide are produced they are cut into wafers such that the exposed face of these wafers is either the crystallographic {100} or {111} planes. The relative structure of these surfaces are important with respect to oxidation, etching and thin film growth. These processes are orientation-sensitive; that is, they depend on the direction in which the crystal slice is cut.

Atom density and dangling bonds

The principle planes in a crystal may be differentiated in a number of ways, however, the atom and/or bond density are useful in predicting much of the chemistry of semiconductor surfaces. Since both silicon and gallium arsenide are *fcc* structures and the {100} and {111} are the only technologically relevant surfaces, discussions will be limited to *fcc* {100} and {111}.

The *atom density* of a surface may be defined as the number of atoms per unit area. Figure 2.8 shows a schematic view of the {111} and {100} planes in a *fcc* lattice. The {111} plane consists of a hexagonal close packed array in which the crystal directions within the plane are oriented at 60° to each other. The hexagonal packing and the orientation of the crystal directions are indicated in Figure 2.8b as an overlaid hexagon.

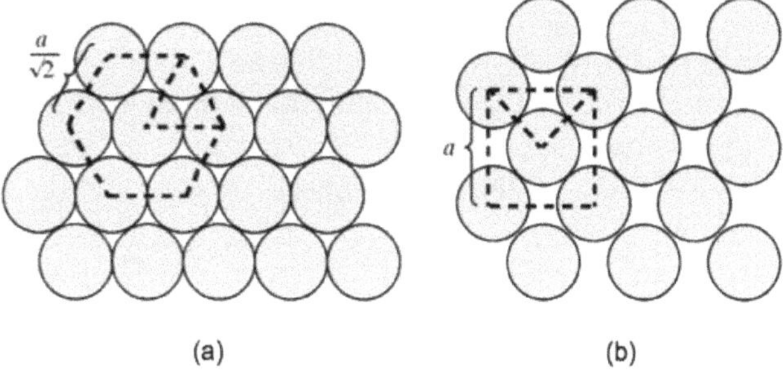

(a)   (b)

Figure 2.8: Schematic representation of (a) the {111} and (b) the {100} faces of a face centered cubic (fcc) lattice, showing the relationship between the close packed rows.

Given the intra-planar inter-atomic distance may be defined as a function of the lattice parameter, the area of this hexagon may be readily calculated. For example in the case of silicon, the hexagon has an area of 38.30 Å^2. The number of atoms within the hexagon is three: the atom in the center plus $^1/_3$ of each of the six atoms at the vertices of the hexagon (each of the atoms at the hexagons vertices is shared by three other adjacent hexagons). Thus, the atom density of the {111} plane is calculated to be 0.0783 Å^{-2}. Similarly, the atom density of the {100} plane may be calculated. The {100} plane consists of a square array in which the crystal directions within the plane are oriented at 90° to each other. Since the square is coincident with one of the faces of the unit cell the area of the square may be readily calculated. For example in the case of silicon, the square has an area of 29.49 Å^2. The number of atoms within the square is 2: the atom in the center plus $^1/_4$ of each of the four atoms at the vertices of the square (each of the atoms at the corners of the square are shared by four other adjacent squares). Thus, the atom density of the $\{100\}$ plane is calculated to be 0.0678 Å^{-2}. While these values for the atom density are specific for silicon, their ratio is constant for all diamond cubic and zinc blende structures: $\{100\}:\{111\} = 1:1.155$. In general, the fewer dangling bonds the more stable a surface structure.

An atom inside a crystal of any material will have a coordination number (n) determined by the structure of the material. For example, all atoms within the bulk of a silicon crystal will be in a tetrahedral four-coordinate environment (n = 4). However, at the surface of a crystal the atoms will not make their full compliment of bonds. Each atom will therefore have less nearest neighbors than an atom within the bulk of the material. The missing bonds are commonly called dangling bonds. While this description is not particularly accurate it is, however, widely employed and as such will be used herein. The number of dangling bonds may be defined as the difference between the ideal coordination number (determined by the bulk crystal structure) and the actual coordination number as observed at the surface.

Figure 2.9 shows a section of the $\{111\}$ surfaces of a diamond cubic lattice viewed perpendicular to the $\{111\}$ plane. The atoms within the bulk have a coordination number of four. In contrast, the atoms at the surface (e.g., the atom shown in blue in Figure 2.9) are each bonded to just three other atoms (the atoms shown in red in Figure 2.9), thus each surface atom has one dangling bond. As can be seen from Figure 2.10, which shows the atoms at the $\{100\}$ surface viewed perpendicular to the $\{100\}$ plane, each atom at the surface (e.g., the atom shown in blue in Figure 2.10) is only coordinated to two other atoms (the atoms shown in red in Figure 2.10), leaving two dangling bonds per atom. It should be noted that the same number of dangling bonds are found for the $\{111\}$ and $\{100\}$ planes of a zinc blende lattice. The ratio of dangling bonds for the $\{100\}$ and $\{111\}$ planes of all diamond cubic and zinc blende structures is $\{100\}:\{111\} = 2:1$. Furthermore, since the atom densities of each plane are known then the ratio of the dangling bond densities is determined to be: $\{100\}:\{111\} = 1:0.577$.

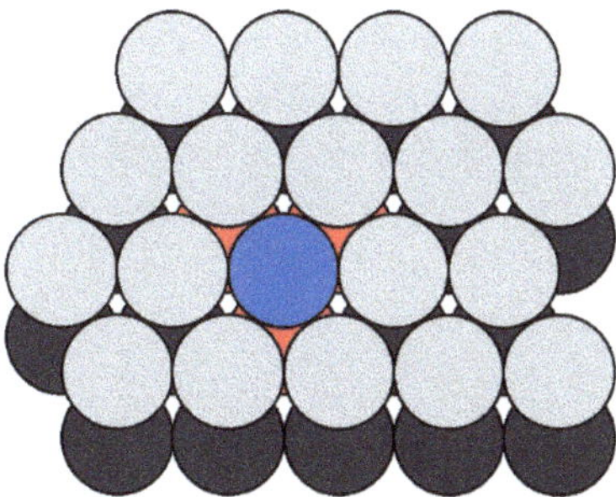

Figure 2.9: A section of the {111} surfaces of a diamond cubic lattice viewed perpendicular to the {111} plane.

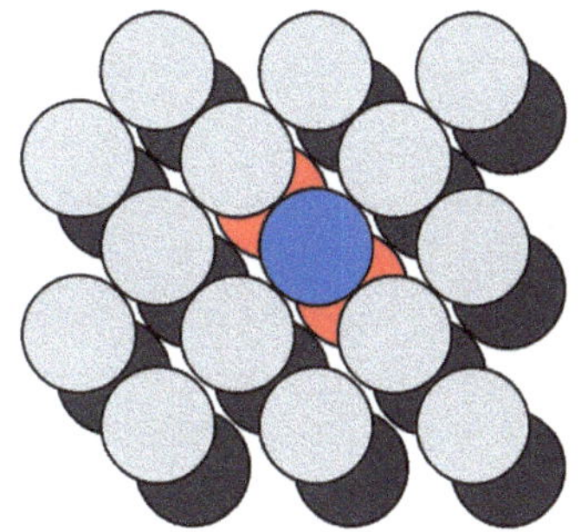

Figure 2.10: A section of the {100} surface of a diamond cubic lattice viewed perpendicular to the {100} plane.

Silicon

For silicon, the {111} planes are closer packed than the {100} planes. As a result, growth of a silicon crystal is therefore slowest in the <111> direction, since it requires laying down a close packed atomic layer upon another layer in its closest packed form. As a consequence <111> Si is the easiest to grow, and therefore the least expensive.

The dissolution or etching of a crystal is related to the number of broken bonds already present at the surface: the fewer bonds to be broken in order to remove an individual atom from a crystal, the easier it will be to dissolve the crystal. As a consequence of having only one dangling bond (requiring three bonds to be broken) etching silicon is slowest in the <111> direction. The electronic properties of a silicon wafer are also related to the number of dangling bonds.

Silicon microcircuits are generally formed on a single crystal *wafer* that is diced after fabrication by either sawing part way through the wafer thickness or scoring (scribing) the surface, and then physically breaking. The physical breakage of the wafer occurs along the natural cleavage planes, which in the case of silicon are the {111} planes.

Gallium arsenide

The zinc blende lattice observed for gallium arsenide results in additional considerations over that of silicon. Although the {100} plane of GaAs is structurally similar to that of silicon, two possibilities exist: a face consisting of either all gallium atoms or all arsenic atoms. In either case the surface atoms have two dangling bonds, and the properties of the face are independent of whether the face is gallium or arsenic.

The {111} plane also has the possibility of consisting of all gallium or all arsenic. However, unlike the {100} planes there is a significant difference between the two possibilities. Figure 2.11 shows the gallium arsenide structure represented by two interpenetrating *fcc* lattices. The [111] axis is vertical within the plane of the page. Although the structure consists of alternate layers of gallium and arsenic stacked along the [111] axis, the distance between the successive layers alternates between large and small.

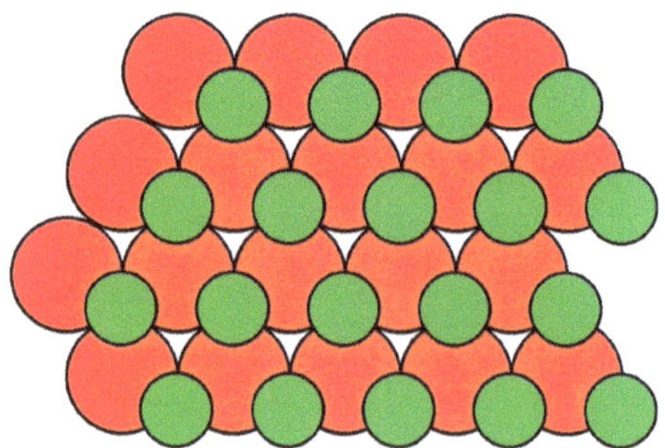

Figure 2.11: The {111} Ga face of GaAs showing a surface layer containing gallium atoms (green) with one dangling bond per gallium and three bonds to the arsenic atoms (red) in the lower layer.

Assigning arsenic as the parent lattice the order of the layers in the [111] direction is As-Ga-As-Ga-As-Ga, while in the [$\bar{1}\bar{1}\bar{1}$] direction the layers are ordered, Ga-As-Ga-As-Ga-As (Figure 2.11). In silicon these two directions are of course identical. The surface of a crystal would be either arsenic, with three dangling bonds, or gallium, with one dangling bond. Clearly, the latter is energetically more favorable. Thus, the (111) plane shown in Figure 2.11 is called the (111) Ga face. Conversely, the [$\bar{1}\bar{1}\bar{1}$] plane would be either gallium, with three dangling bonds, or arsenic, with one dangling bond. Again, the latter is energetically more favorable and the [$\bar{1}\bar{1}\bar{1}$] plane is therefore called the (111) As face.

The (111) As is distinct from that of (111) Ga due to the difference in the number of electrons at the surface. As a consequence, the (111) As face etches more rapidly than the (111) Ga face. In addition, surface evaporation below 770 °C occurs more rapidly at the (111) As face.

Defects in crystalline solids

Up to this point we have only been concerned with ideal structures for crystalline solids in which each atom occupies a designated point in the crystal lattice. Unfortunately, defects ordinarily exist in equilibrium between the crystal lattice and its environment. These *defects* are of two general types: *point defects* and *extended defects*. As their names imply, point defects are associated with a single crystal lattice site, while extended defects occur over a greater range.

Point defects: too many or too few or just plain wrong

Point defects have a significant effect on the properties of a semiconductor, so it is important to understand the classes of point defects and the characteristics of each type. Figure 2.12 summarizes various classes of native point defects; however, they may be divided into two general classes; defects with the wrong number of atoms (deficiency or surplus) and defects where the identity of the atoms is incorrect.

Interstitial impurity

An interstitial impurity occurs when an extra atom is positioned in a lattice site that should be vacant in an ideal structure (Figure 2.12b). Since all the adjacent lattice sites are filled the additional atom will have to squeeze itself into the interstitial site, resulting in distortion of the lattice and alteration in the local behavior of the structure. Small atoms, such as carbon, will prefer to occupy these interstitial sites. Interstitial impurities readily diffuse through the lattice via interstitial diffusion, which can result in a change of the properties of a material as a function of time. Oxygen impurities in silicon generally are located as interstitials.

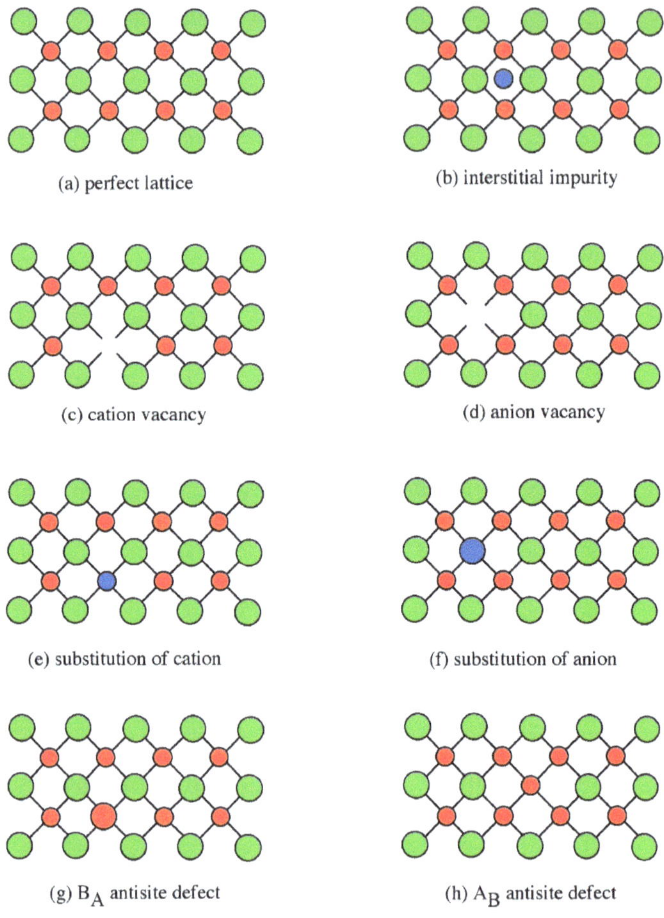

Figure 2.12: Point defects in a crystal lattice.

Vacancies

The converse of an interstitial impurity is when there are not enough atoms in a particular area of the lattice. These are called vacancies. Vacancies exist in any material above absolute zero and increase in concentration with temperature. In the case of compound semiconductors, vacancies can be either cation vacancies (Figure 2.12c) or anion vacancies (Figure 2.12d), depending on what type of atom is missing.

Substitution

Substitution of various atoms into the normal lattice structure is common and used to change the electronic properties of both compound and elemental semiconductors. Any impurity element that is incorporated during crystal growth can occupy a lattice site. Depending on the impurity, substitution defects can greatly distort the lattice and/or alter the electronic structure. In general, cations will try to occupy cation lattice sites (Figure 2.12e), and anion will occupy the anion site (Figure 2.12f). For example, a zinc impurity in GaAs will occupy a gallium site if possible, while sulfur, selenium and tellurium atoms would all try to substitute for arsenic. Some impurities will occupy either site indiscriminately, e.g., Si and Sn occupy both Ga and As sites in GaAs.

Antisite defects

Antisite defects are a particular form of substitution defect and are unique to compound semiconductors. An antisite defect occurs when a cation is misplaced on an anion lattice site or vice versa (Figure 2.12g and h). Dependent on the arrangement these are designated as either A_B antisite defects or B_A antisite defects. For example, an arsenic atom on a gallium lattice site is an As_{Ga} defect. Antisite defects involve fitting lattice site atoms of a different size, than the rest of the lattice, and therefore this often results in a localized distortion of the lattice. In addition, cations and anions will have a different number of electrons in their valence shells, so this substitution will alter the local electron concentration and the electronic properties of this area of the semiconductor.

Chapter 3: III-V Semiconductors

Unlike silicon which has a multitude of uses outside of the semiconductor industry, gallium's main use is in gallium arsenide (GaAs) and related compounds that can convert electricity directly into coherent light (laser diodes) and is employed in electroluminescent light-emitting diodes (LED's); it is also used for doping other semiconductors and in solid-state devices such as heterojunction bipolar transistors (HBTs) and high power high speed metal semiconductor field effect transistors (MESFETs).

The compounds which gallium forms with the Group V (15) elements, GaE (E = N, P, As, Sb) are isoelectronic with the Group 14 elements. There has been considerable interest, particularly in the physical properties of these compounds, since 1952 when Welker first showed that they had semiconducting properties analogous to those of silicon and germanium.

Gallium phosphide, arsenide, and antimonide can all be prepared by direct reaction of the elements; this is normally done in sealed silica tubes or in a graphite crucible under hydrogen. Phase diagram data is hard to obtain in the gallium-phosphorus system because of loss of phosphorus from the bulk material at elevated temperatures. Thus, GaP has a vapor pressure of more than 13.5 atm at its melting point; as compared to 0.89 atm for GaAs. The physical properties of these three compounds are compared with those of the nitride in Table 3.1. All three adopt the zinc blende crystal structure and are more highly conducting than gallium nitride.

Property	GaN	GaP	GaAs	GaSb
Melting point ($°C$)	> 1250 (dec)	1350	1240	712
Density (g/cm^3)	ca. 6.1	4.138	5.3176	5.6137
Crystal structure	Wurtzite	Zinc blende	Zinc blende	Zinc blende
Cell dimen. ($Å$)[a]	$a = 3.187, c = 5.186$	$a = 5.4505$	$a = 5.6532$	$a = 6.0959$
Refractive index[b]	2.35	3.178	3.666	4.388

k (ohm^{-1}cm^{-1})	10^{-9}-10^{-7}	10^{-2}-10^{2}	10^{-6}-10^{3}	6-13
Band gap (eV)^c	3.44	2.24	1.424	0.71

Table 3.1: Physical properties of 13-15 compound semiconductors. [a] Values given for 300 K. [b] Dependent on photon energy; values given for 1.5 eV incident photons. [c] Dependent on temperature; values given for 300 K.

Gallium arsenide versus silicon

Gallium arsenide is a compound semiconductor with a combination of physical properties that has made it an attractive candidate for many electronic applications. From a comparison of various physical and electronic properties of GaAs with those of Si (Table 3.2) the advantages of GaAs over Si can be readily ascertained. Unfortunately, the many desirable properties of gallium arsenide are offset to a great extent by a number of undesirable properties, which have limited the applications of GaAs based devices to date.

Properties	GaAs	Si
Formula weight	144.63	28.09
Crystal structure	Zinc blende	Diamond
Lattice constant	5.6532	5.43095
Melting point (°C)	1238	1415
Density (g/cm^3)	5.32	2.328
Thermal conductivity (W/cm.K)	0.46	1.5
Band gap (eV) at 300 K	1.424	1.12
Intrinsic carrier conc. (cm^{-3})	1.79×10^6	1.45×10^{10}
Intrinsic resistivity (ohm.cm)	10^8	2.3×10^5
Breakdown field (V/cm)	4×10^5	3×10^5
Minority carrier lifetime (s)	10^{-8}	2.5×10^{-3}
Mobility (cm^2/V.s)	8500	1500

Table 3.2: Comparison of physical and semiconductor properties of gallium arsenide and silicon.

Band gap

The band gap of GaAs is 1.42 eV; resulting in photon emission in the infrared range. Alloying GaAs with Al to give $Al_xGa_{1-x}As$ can extend the band gap into the visible red range. Unlike Si, the band gap of GaAs is *direct*, i.e., the transition between the valence band maximum and conduction band minimum involves no momentum change and hence does not require a collaborative particle interaction to occur. Photon generation by *inter-band radiative recombination* is therefore possible in GaAs, whereas in Si, with an indirect band-gap, this process is too inefficient to be of use. The ability to convert electrical energy into light forms the basis of the use of GaAs, and its alloys, in optoelectronics; for example in light emitting diodes (LEDs), solid state lasers (light amplification by the stimulated emission of radiation).

A significant drawback of small band gap semiconductors, such as Si, is that electrons may be thermally promoted from the valence band to the conduction band. Thus, with increasing temperature the thermal generation of carriers eventually becomes dominant over the intentionally doped level of carriers. The wider band gap of GaAs gives it the ability to remain 'intentionally' semiconducting at higher temperatures; GaAs devices are generally more stable to high temperatures than a similar Si device.

Carrier density

The low intrinsic carrier density of GaAs in a pure (undoped) form indicates that GaAs is intrinsically a very poor conductor and is commonly referred to as being *semi-insulating*. Since carrier density is usually altered by adding dopants of either the p- (positive) or n- (negative) type, the semi-insulating property of GaAs allows many active devices to be grown on a single substrate, where the semi-insulating GaAs provides the electrical isolation of each device; an important feature in the miniaturization of electronic circuitry, i.e., VLSI (very-large-scale-integration) involving over 100,000 components per chip (one chip is typically between 1 and 10 mm square).

Electron mobility

The higher electron mobility in GaAs than in Si potentially means that in devices where electron transit time is the critical performance parameter, GaAs devices will operate with higher response times than equivalent Si devices. However, the fact that hole mobility is similar for both GaAs and Si means that devices relying on cooperative electron and hole movement, or hole movement alone, show no improvement in response time when GaAs based.

Crystal growth

The bulk crystal growth of GaAs presents a problem of stoichiometric control due the loss, by evaporation, of arsenic both in the melt and the growing crystal (>600 °C). Melt growth techniques are, therefore, designed to enable an overpressure of arsenic above the melt to be maintained, thus preventing evaporative losses. The loss of arsenic also negates diffusion techniques commonly used for wafer doping in Si technology, since the diffusion temperatures required exceed that of arsenic loss.

Crystal stress

The thermal gradient and, hence, stress generated in melt grown crystals have traditionally limited the maximum diameter of GaAs wafers because with increased wafer diameters the thermal stress generated dislocation (crystal imperfections) densities eventually becomes unacceptable for device applications. Until recently 5.9" (150 mm) diameter wafer of GaAs were more common that the 11.8" (300 mm) used for Si, however, advances in crystal growth and wafer fabrication have meant GaAs wafers of sizes comparable to those of Si are now available.

Physical strength

Gallium arsenide single crystals are very brittle, requiring that considerably thicker substrates than those employed for Si devices.

Native oxide

Gallium arsenide's native oxide is found to be a mixture of non-stoichiometric gallium and arsenic oxides and elemental arsenic. Thus, the electronic band structure is severely disrupted, causing a breakdown in 'normal' semiconductor behavior on the GaAs surface. As a consequence, the GaAs MISFET (metal-insulator-semiconductor-field-effect-transistor) equivalent to the technologically important Si based MOSFET (metal-oxide-semiconductor-field-effect-transistor) is, therefore, presently unavailable. The passivation of the surface of GaAs is therefore a key issue when endeavoring to utilize the FET technology using GaAs. Passivation in this discussion means the reduction in mid-gap band states, which destroy the semiconducting properties of the material. Additionally, this also means the production of a chemically inert coating, which prevents the formation of additional reactive states, which can affect the properties of the device.

Chapter 4: Synthesis and Purification of Bulk Semi-conductors

The synthesis and purification of bulk polycrystalline semiconductor material represents the first step towards the commercial fabrication of an electronic device. This polycrystalline material is then used as the raw material for the formation of single crystal material that is processed to semiconductor wafers. The strong influence on the electric characteristics of a semiconductors exhibited by small amounts of some impurities requires that the bulk raw material be of very high purity (>99.9999%). Although some level of purification is possible during the crystallization process it is important to use as high a purity starting material as possible. While a wide range of substrate materials are available from commercial vendors, silicon and GaAs represent the only large-scale commercial semiconductor substrates, and thus the discussion will be limited to the synthesis and purification of these materials.

Semiconductor grade silicon

The synthesis and purification of bulk polycrystalline semiconductor material represents the first step towards the commercial fabrication of an electronic device. This polycrystalline material is then used as the raw material for the formation of single crystal material that is processed to semiconductor wafers. The strong influence on the electric characteristics of a semiconductors exhibited by small amounts of some impurities requires that the bulk raw material be of very high purity (>99.9999%). Although some level of purification is possible during the crystallization process it is important to use as high a purity starting material as possible.

While 98% elemental silicon, known as *metallurgical-grade* silicon (MGS), is produced on a large scale, the requirements of extreme purity for electronic device fabrication require additional purification steps in order to produce *electronic-grade* silicon (EGS). In order for the purity levels to be acceptable for subsequent crystal growth and device fabrication, EGS must have carbon and oxygen impurity levels less than a few parts per million (ppm), and metal impurities at the parts per billion (ppb) range or lower. Table 4.1 and Table 4.2 give typical impurity concentrations in MGS and EGS, respectively.

Element	Concentration (ppm)
Aluminum	1000-4350
Boron	40-60
Calcium	245-500
Chromium	50-200
Copper	15-45
Iron	1550-6500
Magnesium	10-50
Manganese	50-120
Molybdenum	< 20
Nickel	10-105
Phosphorus	20-50
Titanium	140-300
Vanadium	50-250
Zirconium	20

Table 4.1: Typical impurity concentrations found in metallurgical-grade silicon (MGS).

Element	Concentration (ppb)
Arsenic	< 0.001
Antimony	< 0.001
Boron	≤ 0.1
Carbon	100-1000
Chromium	< 0.01
Cobalt	0.001
Copper	0.1
Gold	< 0.00001
Iron	0.1-1.0
Nickel	0.1-0.5
Oxygen	100-400
Phosphorus	≤ 0.3
Silver	0.001
Zinc	< 0.1

Table 4.2: Typical impurity concentrations found in electronic-grade silicon (EGS).

Metallurgical-grade silicon (MGS)

The typical source material for commercial production of elemental silicon is quartzite gravel; a relatively pure form of sand (SiO_2). The first step in the synthesis of silicon is the melting and reduction of the silica in a submerged-electrode arc furnace. An example of which is shown schematically in Figure 4.1, along with the appropriate chemical reactions. A mixture of quartzite gravel and carbon are heated to high temperatures (ca. 1800 °C) in the furnace. The carbon bed consists of a mixture of coal, coke, and wood chips. The latter providing the necessary porosity such that the gases created during the reaction (SiO and CO) are able to flow through the bed.

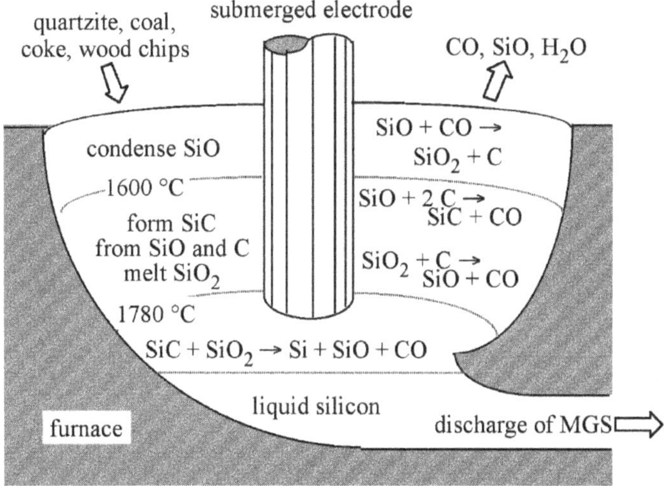

Figure 4.1: Schematic of submerged-electrode arc furnace for the production of metallurgical-grade silicon (MGS).

The overall reduction reaction of SiO_2 is expressed,

$$SiO_2(liquid) + 2\ C(solid) \rightarrow Si(liquid) + 2\ CO\ (gas)$$

however, the reaction sequence is more complex than this overall reaction implies and involves the formation of SiC and SiO intermediates. The initial reaction between molten SiO_2 and C takes place in the arc between adjacent electrodes, where the local temperature can exceed 2000 °C.

$$SiO_2 + 2\,C \;\underset{<1600\,°C}{\overset{>1700\,°C}{\rightleftharpoons}}\; SiO + CO$$

The SiO and CO thus generated flow to cooler zones in the furnace where SiC is formed, or higher in the bed where they reform SiO_2 and C:

$$SiO + 2\,C \;\rightarrow\; SiC + CO \quad (1600 - 1700\,°C)$$

The SiC reacts with molten SiO_2, producing the desired silicon along with SiO and CO:

$$SiC + SiO_2 \rightarrow Si + SiO + CO$$

The molten silicon formed is drawn-off from the furnace and solidified.

The as-produced MGS is approximately 98-99% pure, with the major impurities being aluminum and iron (Table 4.1), however, obtaining low levels of boron impurities is of particular importance, because it is difficult to remove and serves as a dopant for silicon. The drawbacks of the above process are that it is both energy and raw material intensive. It is estimated that the production of one metric ton (1,000 kg) of MGS requires 2500-2700 kg quartzite, 600 kg charcoal, 600-700 kg coal or coke, 300-500 kg wood chips, and 500,000 kWh of electric power. Currently, approximately 500,000 metric tons of MGS are produced per year, worldwide. Most of the production (ca. 70%) is used for metallurgical applications (e.g., aluminum-silicon alloys are commonly used for automotive engine blocks) from whence its name is derived. Applications in a variety of chemical products such as silicone resins account for about 30%, and only 1% or less of the total production of MGS is used in the manufacturing of high-purity EGS for the electronics industry. The current worldwide consumption of EGS is approximately 5×10^6 kg per year.

Electronic-grade silicon (EGS)

Electronic-grade silicon (EGS) is a polycrystalline material of exceptionally high purity and is the raw material for the growth of single-crystal silicon. EGS is one of the purest materials commonly available (Table 4.2). The formation of EGS from MGS is accomplished through chemical purification processes. The basic concept of which involves the conversion of MGS to a volatile silicon compound, which is purified by distillation, and subsequently decomposed to re-form elemental silicon of higher purity (i.e., EGS). Irrespective of the purification route employed, the first step is physical pulverization of MGS followed by its conversion to the volatile silicon compounds.

A number of compounds, such as monosilane (SiH_4, Figure 4.2), dichlorosilane (SiH_2Cl_2, Figure 4.3), trichlorosilane ($SiHCl_3$, Figure 4.4), and silicon tetrachloride ($SiCl_4$, Figure 4.5), have been considered as chemical intermediates. Among these, $SiHCl_3$ has been used predominantly as the intermediate compound for subsequent EGS formation, although SiH_4 is used to a lesser extent. Silicon tetrachloride and its lower chlorinated derivatives are used for the chemical vapor deposition (CVD) growth of Si and SiO_2. The boiling points of silane and its chlorinated products (Table 5.3) are such that they are conveniently separated from each other by fractional distillation.

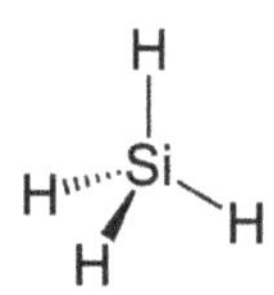

Figure 4.2: Structure of monosilane (SiH_4).

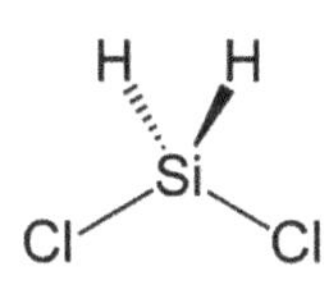

Figure 4.3: Structure of dichlorosilane (SiH_2Cl_2).

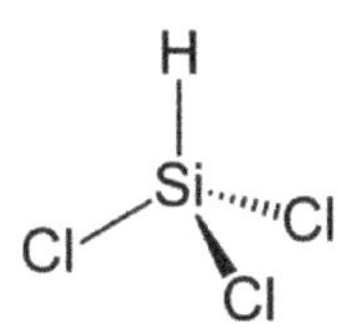

Figure 4.4: Structure of trichlorosilane ($SiHCl_3$).

$$
\begin{array}{c}
Cl \\
| \\
Cl-Si_{\cdots\cdots}Cl \\
| \\
Cl
\end{array}
$$

Figure 4.5: silicon tetrachloride ($SiCl_4$).

Compound	Boiling point (°C)
SiH_4	-112.3
SiH_3Cl	-30.4
SiH_2Cl_2	8.3
$SiHCl_3$	31.5
$SiCl_4$	57.6

Table 4.3: Boiling points of silane and chlorosilanes at 760 mmHg (1 atmosphere).

The reasons for the predominant use of $SiHCl_3$ in the synthesis of EGS are as follows:

- $SiHCl_3$ can be easily formed by the reaction of anhydrous hydrogen chloride with MGS at reasonably low temperatures (200-400 °C);
- it is liquid at room temperature (bp = 31.5 °C) so that purification can be accomplished using standard distillation techniques;
- it is easily handled and if dry can be stored in carbon steel tanks;
- its liquid is easily vaporized and, when mixed with hydrogen it can be transported in steel lines without corrosion;
- it can be reduced at atmospheric pressure in the presence of hydrogen;
- its deposition can take place on heated silicon, thus eliminating contact with any foreign surfaces that may contaminate the resulting silicon; and
- it reacts at lower temperatures (1000-1200 °C) and at faster rates than does $SiCl_4$.

Chlorosilane (Seimens) process

Trichlorosilane ($SiHCl_3$, Figure 4.4) is synthesized by heating powdered MGS with anhydrous hydrogen chloride (HCl) at around 300 °C in a fluidized-bed reactor,

$$Si(solid) + 3\ HCl(gas) \underset{>900\ °C}{\overset{ca.\ 300\ °C}{\rightleftharpoons}} SiHCl_3(vapor) + H_2\ (gas)$$

Since the reaction is actually in equilibrium and the formation of $SiHCl_3$ highly *exothermic*, efficient removal of generated heat is essential to assure a maximum yield of $SiHCl_3$. While the stoichiometric reaction is that shown above, a mixture of chlorinated silanes is actually prepared which must be separated by fractional distillation, along with the chlorides of any impurities. In particular iron, aluminum, and boron are removed as $FeCl_3$ (boiling point = 316 °C), $AlCl_3$ (melting point = 190 °C, sublimes), and BCl_3 (boiling point = 12.65 °C), respectively. Fractional distillation of $SiHCl_3$ from these impurity halides result in greatly increased purity with a concentration of electrically active impurities of less than 1 ppb.

EGS is prepared from purified $SiHCl_3$ in a chemical vapor deposition (CVD) process similar to the epitaxial growth of Si. The high-purity $SiHCl_3$ is vaporized, diluted with high-purity hydrogen, and introduced into the Seimens deposition reactor, shown schematically in Figure 5.6. Inside the reactor thin (4 mm in diameter) silicon rods, called *slim rods*, are supported by graphite electrodes. Resistance heating of the slim rods causes the decomposition of the $SiHCl_3$ to yield silicon.

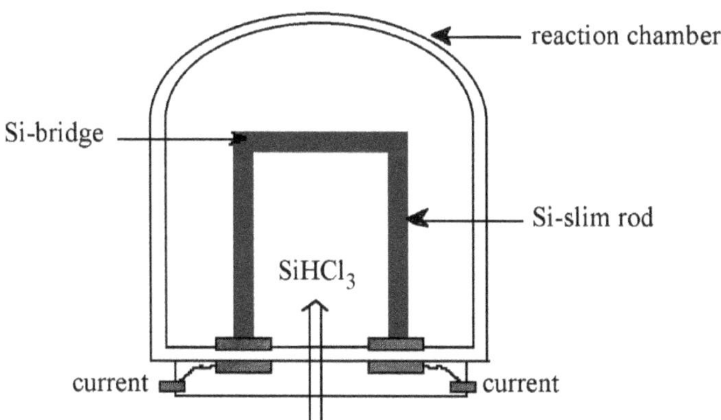

Figure 4.6: Schematic representation of a Seimens deposition reactor.

The shift in the equilibrium from forming $SiHCl_3$ from Si at low temperature, to forming Si from $SiHCl_3$ at high temperature is as a consequence of the temperature (T) dependence,

$$\ln K_p = \frac{-\Delta H}{RT}$$

of the equilibrium constant (K_p), where ρ = partial pressure.

$$K_p = \frac{\rho_{SiHCl_3} \; \rho_{H_2}}{\rho_{HCl}}$$

Since the formation of SiHCl$_3$ is *exothermic*, i.e., $\Delta H < 0$, an increase in the temperature causes the partial pressure of SiHCl$_3$ to decrease. Thus, the Siemens process is typically run at ca. 1100 °C, while the reverse fluidized bed process is carried out at 300 °C.

The slim rods act as a nucleation point for the deposition of silicon, and the resulting polycrystalline rod consists of columnar grains of silicon (polysilicon) grown perpendicular to the rod axis. Growth occurs at less than 1 mm per hour, and after deposition for 200 to 300 hours high-purity (EGS) polysilicon rods of 150 - 200 mm in diameter are produced. For subsequent oatzone refining the polysilicon EGS rods are cut into long cylindrical rods. Alternatively, the as-formed polysilicon rods are broken into chunks for single crystal growth processes, for example Czochralski melt growth.

In addition to the formation of silicon, the HCl co-product reacts with the SiHCl$_3$ (Figure 4.4) reactant to form silicon tetrachloride (SiCl$_4$, Figure 4.5) and hydrogen as major byproducts of the process,

$$HCl + SiHCl_3 \rightarrow SiCl_4 + H_2$$

This reaction represents a major disadvantage with the Seimens process: poor efficiency of silicon and chlorine consumption. Typically, only 30% of the silicon introduced into CVD reactor is converted into high-purity polysilicon. In order to improve efficiency, the HCl, SiCl$_4$, H$_2$, and unreacted SiHCl$_3$ are separated and recovered for recycling.

Figure 4.7 illustrates the entire chlorosilane process starting with MGS and including the recycling of the reaction byproducts to achieve high overall process efficiency. As a consequence, the production cost of high-purity

EGS depends on the commercial usefulness of the byproduct, $SiCl_4$. Additional disadvantages of the Seimens process are derived from its relatively small batch size, slow growth rate, and high-power consumption. These issues lead to the investigation of alternative cost-efficient routes to EGS.

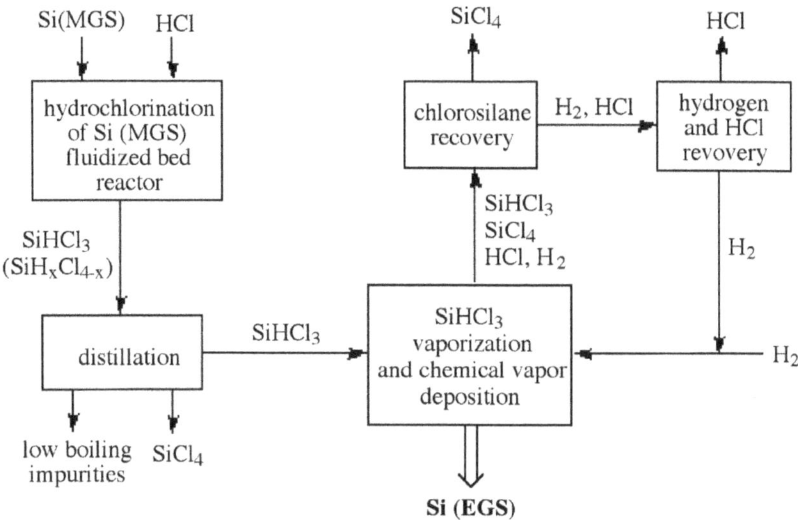

Figure 4.7: Schematic representation of the reaction pathways for the formation of EGS using the chlorosilane process.

Silane process

An alternative process for the production of EGS that has begun to receive commercial attention is the pyrolysis of silane (SiH_4, Figure 4.2). The advantages of producing EGS from SiH_4 instead of $SiHCl_3$ are potentially lower costs associated with lower reaction temperatures, and less harmful byproducts. Silane decomposes < 900 °C to give silicon and hydrogen,

$$SiH_4(vapor) \rightarrow Si(solid) + 2\,H_2\,(gas)$$

Silane may be prepared by a number of routes, each having advantages with respect to purity and production cost. The simplest process involves the direct reaction of MGS powders with magnesium at 500 °C in a hydrogen atmosphere, to form magnesium silicide (Mg_2Si). The magnesium silicide is then reacted with ammonium chloride in liquid ammonia below 0 °C,

$$Mg_2Si + 4\,NH_4Cl \;\rightarrow\; SiH_4 + 2\,MgCl_2 + 5\,NH_3$$

This process is ideally suited to the removal of boron impurities (a p-type dopant in Si), because the diborane (B_2H_6) produced during the reaction forms the Lewis acid-base complex, $H_3B(NH_3)$, whose volatility is sufficiently lower than SiH_4, allowing for the purification of the latter. It is possible to prepare EGS with a boron content of ≤ 20 ppt using SiH_4 synthesized in this manner. However, phosphorus (another dopant) in the form of PH_3 may be present as a contaminant requiring subsequent purification of the SiH_4.

Alternative routes to SiH_4 involve the chemical reduction of $SiCl_4$ by either lithium hydride (LiH),

$$SiCl_4 + 4\,LiH \;\rightarrow\; SiH_4 + 4\,LiCl$$

lithium aluminum hydride ($LiAlH_4$),

$$SiCl_4 + 4\,LiAlH_4 \;\rightarrow\; SiH_4 + LiCl + AlCl_3$$

or via hydrogenation in the presence of elemental silicon.

$$SiCl_4 + 2\,H_2 + Si(98\%) \;\rightarrow\; 4\,SiHCl_3$$

$$2\,SiHCl_3 \;\rightarrow\; SiH_2Cl_2 + SiCl_4$$

$$3\,SiH_2Cl_2 \;\rightarrow\; SiH_3Cl + 2\,SiHCl_3$$

$$2\,SiH_3Cl \;\rightarrow\; SiH_4 + SiH_2Cl_2$$

The hydride reduction reactions may be carried-out on relatively large scales (ca. 50 kg), but only batch processes. In contrast, Union Carbide has adapted the hydrogenation to a continuous process, involving disproportionation reactions of chlorosilanes and the fractional distillation of silane, Table 4.3.

Pyrolysis of silane on resistively heated polysilicon laments at 700 – 800 °C yields polycrystalline EGS. As noted above, the EGS formed has remarkably low boron impurities compared with material prepared from trichlorosilane. Moreover, the resulting EGS is less contaminated with transition metals from the reactor container because SiH_4 decomposition does not cause as much of a corrosion problem as is observed with halide precursor compounds.

Granular polysilicon deposition

Both the chlorosilane (Seimens) and silane processes result in the formation of rods of EGS. However, there has been increased interest in the formation of granular polycrystalline EGS. This process was developed in 1980's and relies on the decomposition of SiH_4 in a fluidized-bed deposition reactor to produce free-flowing granular polysilicon.

Tiny silicon particles are fluidized in a SiH_4/H_2 flow, and act as seed crystal onto which polysilicon deposits to form free-flowing spherical particles. The size distribution of the particles thus formed is over the range from 0.1 to 1.5 mm in diameter with an average particle size of 0.7 mm. The fluidized-bed, seed particles are originally made by grinding EGS in a ball (or hammer) mill and leaching the product with acid, hydrogen peroxide, and water. This process is time-consuming and costly, and tended to introduce undesirable impurities from the metal grinders. In an alternative method, large EGS particles are fired at each other, by a high-speed stream of an inert gas, and the collision breaks them down into particles of suitable size for a fluidized bed. This process has the main advantage that it introduces no foreign materials and requires no leaching or other post purification.

The fluidized-bed reactors are much more efficient than traditional rod reactors as a consequence of the greater surface area available during CVD growth of silicon. It has been suggested that fluidized-bed reactors require $^1/_5$ to $^1/_{10}$ the energy, and half the capital cost of the traditional process. The quality of fluidized-bed polysilicon has proven to be equivalent to polysilicon produced by the conventional methods. Moreover, granular EGS in a free- owing form, and with high bulk density, enables crystal growers to obtain the high, reproducible production yields out of each crystal growth run. For example, in the Czochralski crystal growth process, crucibles can be quickly and easily filled to uniform loading with granular EGS, which typically exceed those of randomly stacked polysilicon chunks produced by the Siemens silane process.

Zone refining

The technique of zone refining is used to purify solid materials and is commonly employed in metallurgical refining. In the case of silicon may be used to obtain the desired ultimate purity of EGS, which has already been purified by chemical processes. Zone refining makes use of the fact that the equilibrium solubility of any impurity (e.g., Al) is different in the solid and liquid phases of a material (e.g., Si). For the dilute solutions, as is observed in EGS silicon, an equilibrium segregation coefficient (k_0) is defined by

$$k_0 = C_s/C_l$$

where C_s and C_l are the equilibrium concentrations of the impurity in the solid and liquid near the interface, respectively.

If k_0 is less than 1 then the impurities are left in the melt as the molten zone is moved along the material. In a practical sense a molten zone is established in a solid rod. The zone is then moved along the rod, from left to right. If k_0 <1 then the frozen part left on the trailing edge of the moving molten zone will be purer than the material that melts in on the right-side leading edge of the moving molten zone. Consequently, the solid to the left of the molten zone is purer than the solid on the right. At the completion of the first pass the impurities become concentrated to the right of the solid sample. Repetition of the process allows for purification to exceptionally high levels. Table 5.4 lists the equilibrium segregation coefficients for common impurity and dopant elements in silicon; it should be noted that they are all less than 1.

Element	k_0
Aluminum	0.002
Boron	0.8
Carbon	0.07
Copper	4×10^{-6}
Iron	8×10^{-6}
Oxygen	0.25
Phosphorus	0.35
Antimony	0.023

Table 4.4: Segregation coefficients for common impurity and dopant elements in silicon.

Gallium arsenide

In contrast to electronic grade silicon (EGS), whose use is a minor fraction of the global production of elemental silicon gallium arsenide (GaAs) is produced exclusively for use in the semiconductor industry. However, arsenic and its compounds have significant commercial applications. The main use of elemental arsenic is in alloys of Pb, and to a lesser extent Cu, while arsenic compounds are widely used in pesticides and wood preservatives and the production of bottle glass. Thus, the electronics industry represents a minor user of arsenic. In contrast, although gallium has minor uses as a high-temperature liquid seal, manometric fluids and heat transfer media, and for low temperature solders, its main use is in semiconductor technology.

Isolation and purification of gallium metal

At 19 ppm gallium (L. Gallia, France) is about as abundant as nitrogen, lithium and lead; it is twice as abundant as boron (9 ppm), but is more difficult to extract due to the lack of any major gallium-containing ore. Gallium always occurs in association either with zinc or germanium, its neighbors in the periodic table, or with aluminum in the same group. Thus, the highest concentrations (0.1-1%) are in the rare mineral germanite (a complex sulfide of Zn, Cu, Ge, and As), while concentrations in sphalerite (ZnS), diaspore [AlO(OH)], bauxite, or coal, are a hundred-fold less. Industrially, gallium was originally recovered from the flue dust emitted during sulfide roasting or coal burning (up to 1.5% Ga); however, it is now obtained as side product of vast aluminum industry and in particular from the Bayer process for obtaining alumina from bauxite.

The Bayer process involves dissolution of bauxite (AlO_xOH_{3-2x}) in aqueous NaOH, separation of insoluble impurities, partial precipitation of the trihydrate [$Al(OH)_3$] and calcination at 1,200 °C. During processing the alkaline solution is gradually enriched in gallium from an initial weight ratio Ga/Al of about $^1/_{5000}$ to about $^1/_{300}$. Electrolysis of these extracts with an Hg cathode results in further concentration, and the solution of sodium gallate, $Na[Ga(OH)_4]$, thus formed is then electrolyzed with a stainless steel cathode to give Ga metal. Since bauxite contains 0.003-0.01% gallium, complete recovery would yield some 500-1000 tons per annum, however present consumption is only 0.1% of this about 10 tons per annum.

A typical analysis of the 98-99% pure gallium obtained as a side product from the Bayer process is shown in Table 4.5. This material is further purified to 99.99% by chemical treatment with acids and O_2 at high temperatures followed by crystallization. This chemical process results in the reduction of the majority of metal impurities at the ppm level, see Table 4.5.

Element	Bayer process (ppm)	After acid/base leaching (ppm)	500 zone passes (ppm)
Aluminum	100-1,000	7	< 1
Calcium	10-100	Not detected	Not detected
Copper	100-1,000	2	< 1
Iron	100-1,000	7	< 1
Lead	< 2000	30	Not detected
Magnesium	10-100	1	Not detected
Mercury	10-100	Not detected	Not detected
Nickel	10-100	Not detected	Not detected
Silicon	10-100	≈ 1	Not detected
Tin	10-100	≈ 1	Not detected
Titanium	10-100	1	< 1
Zinc	30,000	≈ 1	Not detected

Table 4.5: Typical analysis of gallium obtained as a side product from the Bayer process.

Purification to seven nines 99.9999% is possible through *zone refining*, however, since the equilibrium distribution coefficient of the residual impurities $k_0 \approx 1$, multiple passes are required, typically > 500. The low melting point of gallium ensures that contamination from the container wall (which is significant in silicon zone refining) is minimized. In order to facilitate the multiple zone refining in a suitable time, a simple modification of zone refining is employed shown in Figure 4.8.

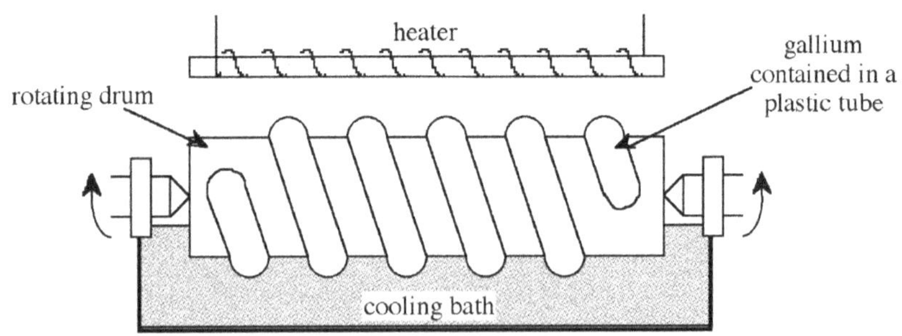

Figure 4.8: Schematic representation of a zone refining apparatus for gallium metal purification.

The gallium is contained in a plastic tube (c.f., Figure 4.8) wrapped around a rotating cylinder that is half immersed in a cooling bath. A heater is positioned above the gallium plastic coil. Thus, establishing a series of molten zones that pass upon rotation of the drum by one helical segment per revolution. In this manner, 500 passes may be made in relatively short time periods. The typical impurity levels of gallium zone refined in this manner are given in Table 4.5.

Isolation and purification of elemental arsenic

Elemental arsenic (L. arsenicum, yellow orpiment) exists in two forms: yellow (cubic, As_4) and gray or metallic (rhombohedral). At a natural abundance of 1.8 ppm arsenic is relatively rare, however, this is offset by its presence in a number of common minerals and the relative ease of isolation. Arsenic containing minerals are grouped into three main classes: the sulfides realgar (As_4S_4) and orpiment (As_2S_3), the oxide arsenolite (As_2O_3), and the arsenides and sulfaresenides of the iron, cobalt, and nickel. Minerals in this latter class include: loellinginite ($FeAs_2$), safforlite (CoAs), niccolite (NiAs), rammelsbergite ($NiAs_2$), ansenopyrite or mispickel (FeAsS), cobaltite (CoAsS), enargite (Cu_3AsS_4), gerdsorfite (NiAsS), and the quarturnary sulfide glaucodot [(Co,Fe)AsS]. Table 4.6 shows the typical impurities in arsenopyrite.

Element	Concentration (ppm)
Silver	90
Gold	8
Cobalt	30,000
Copper	200
Germanium	30
Manganese	3,000
Molybdenum	60
Nickel	< 3,000
Lead	50
Platinum	0.4
Rhenium	50

Selenium	50
Vanadium	300
Zinc	400

Table 4.6: Typical impurities in arsenopyrite (FeAsS).

Arsenic is obtained commercially by smelting either $FeAs_2$ or FeAsS at 650-700 °C in the absence of air and condensing the sublimed element (T_{sub} = 613 °C),

$$\overset{\text{650-700 °C}}{\text{FeAsS} \rightarrow} \text{FeS + As(vapor)} \overset{\text{<613 °C}}{\rightarrow} \text{As(solid)}$$

The arsenic thus obtained is combined with lead and then sublimed (T_{sub} = 614 °C) which binds any sulfur impurities more strongly than arsenic. Residual arsenic that remains trapped in the iron sulfide is separated by roasting the sulfide in air via the formation of the oxide (As_2O_3). The oxide is sublimed into the flue system during roasting from where it is collected and reduced with charcoal at 700-800 °C to give elemental arsenic. Semiconductor grade arsenic (>99.9999%) is formed by zone refining.

Synthesis and purification of gallium arsenide.

Gallium arsenide can be prepared by the direct reaction of the elements,

$$\overset{\text{>1240 °C}}{\text{Ga(liquid) + As(vapor)} \rightarrow} \text{GaAs(solid)}$$

While conceptually simple, however, the synthesis of GaAs is complicated by the different vapor pressures of the reagents and the highly exothermic nature of the reaction. Furthermore, since the synthesis of GaAs at atmospheric pressure is accompanied by its simultaneous decomposes due to the loss by sublimation, of arsenic, the synthesis must be carried out under an overpressure of arsenic in order to maintain a stoichiometric composition of the synthesized GaAs. In order to overcome the problems associated with arsenic loss, the reaction is usually carried out in a sealed reaction tube. However, if a stoichiometric quantity of arsenic is used in the reaction a constant temperature of 1238 °C must be employed in order to maintain the desired arsenic overpressure of 1 atm. Practically, it is easier to use a large excess of arsenic heated to a lower temperature. In this situation the pressure in the tube is approximately equal to the equilibrium vapor pressure of the volatile component (arsenic) at the lower temperature. Thus, an over pressure of 1 atm. arsenic may be maintained if within a sealed tube elemental arsenic is heated to 600-620 °C while the GaAs is maintained at 1240-1250 °C.

Figure 4.9 shows the sealed tube configuration that is typically used for the synthesis of GaAs. The tube is heated within a two-zone furnace. The boats holding the reactants are usually made of quartz, however, graphite is also used since the latter has a closer thermal expansion match to the GaAs product. If higher purity is required then pyrolytic boron nitride (PBN) is used. One of the boats is loaded with pure gallium the other with arsenic. A plug of quartz wool may be placed between the boats to act as a diffuser. The tube is then evacuated and sealed. Once brought to the correct reaction temperatures (Figure 4.9), the arsenic vapor is transported to the gallium, and they react to form GaAs in a controlled manner. Table 4.7 gives the typical impurity concentrations found in polycrystalline GaAs.

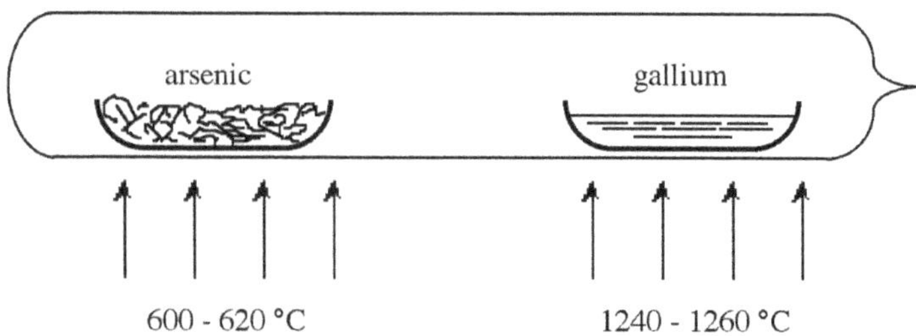

Figure 4.9: Schematic representation of a sealed tube synthesis of GaAs.

Element	Concentration (ppm)
Boron	0.1
Carbon	0.7
Nitrogen	0.1
Oxygen	0.5
Fluorine	0.2
Magnesium	0.02
Aluminum	0.02
Silicon	0.02

Phosphorus	0.1
Sulfur	0.01
Chlorine	0.08
Nickel	0.04
Copper	0.01
Zinc	0.05

Figure 4.7: Impurity concentrations found in polycrystalline GaAs.

Polycrystalline GaAs, formed in from the direct reaction of the elements is often used as the starting material for single crystal growth via Bridgeman or Czochralski crystal growth. It is also possible to prepare single crystals of GaAs directly from the elements using in-situ, or direct, compounding within a high-pressure liquid encapsulated Czochralski (HPLEC) technique.

Growth of gallium arsenide crystals

When considering the synthesis of Group 13-15 compounds for electronic applications, the very nature of semiconductor behavior demands the use of high purity single crystal materials. The polycrystalline materials synthesized above are, therefore, of little use for 13-15 semiconductors but may serve as the starting material for melt grown single crystals. For GaAs grown single crystals are achieved by one of two techniques: the Bridgman technique, and the Czochralski technique.

Bridgman growth

The Bridgman technique requires a two-zone furnace, of the type shown in Figure 4.10. The left hand zone is maintained at a temperature of *ca.* 610 °C, allowing sufficient overpressure of arsenic within the sealed system to prevent arsenic loss from the gallium arsenide. The right hand side of the furnace contains the polycrystalline GaAs raw material held at a temperature just above its melting point (*ca.* 1240 °C). As the furnace moves from left to right, the melt cools and solidifies. If a seed crystal is placed at the left hand side of the melt (at a point where the temperature gradient is such that only the end melts), a specific orientation of single crystal may be propagated at the liquid-solid interface eventually to produce a single crystal.

Czochralski growth

The Czochralski technique, which is the most commonly used technique in industry, is shown in Figure 4.11. The process relies on the controlled withdrawal of a *seed crystal* from a liquid melt. As the seed is lowered into the melt, partial melting of the tip occurs creating the liquid solid interface required for crystal growth. As the seed is withdrawn, solidification occurs and the seed orientation is propagated into the grown material. The variable parameters of rate of withdrawal and rotation rate can control crystal diameter and purity. As shown in Figure 4.11 the GaAs melt is capped by boron trioxide (B_2O_3). The capping layer, which is inert to GaAs, prevents arsenic loss when the pressure on the surface is above atmospheric pressure. The growth of GaAs by this technique is thus termed liquid encapsulated Czochralski (LEC) growth.

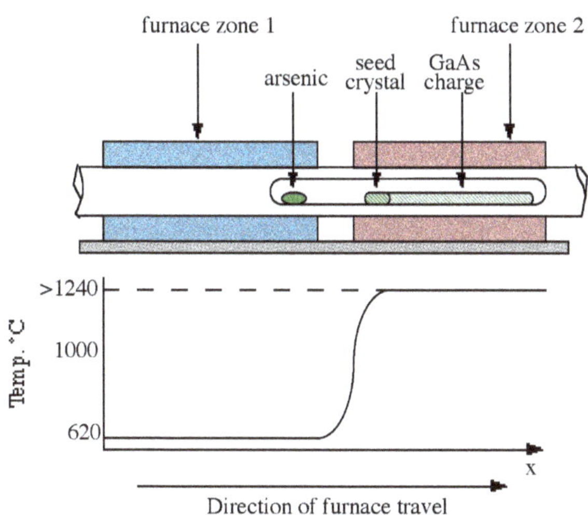

Figure 4.10: A schematic diagram of a Bridgman two-zone furnace used for melt growths of single crystal GaAs.

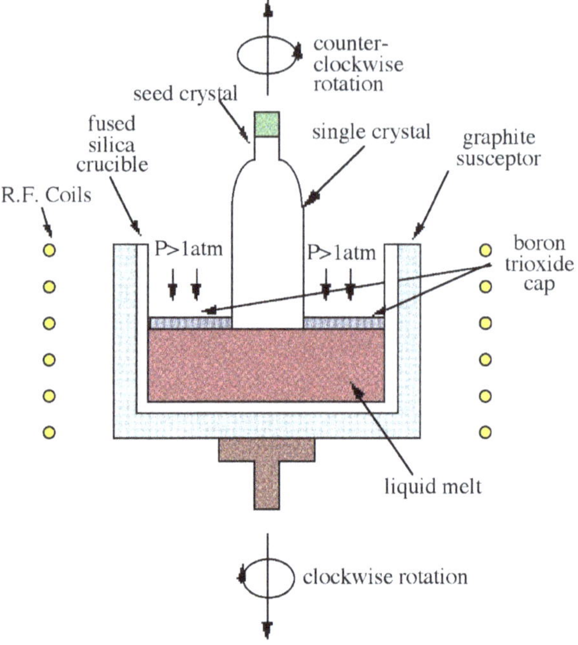

Figure 4.11: A schematic diagram of the Czochralski technique as used for growth of GaAs single crystal bond.

While the Bridgman technique is largely favored for GaAs growth, larger diameter wafers can be obtained by the Czochralski method. Both of these melt techniques produce materials heavily contaminated by the crucible, making them suitable almost exclusively as substrate material. Another disadvantage of these techniques is the production of defects in the material caused by the melt process.

Chapter 5: Wafer Formation and Processing

Integrated circuits (ICs) and discrete solid-state devices are manufactured on semiconductor wafers. Silicon based devices are made on silicon wafers, while III-V (13-15) semiconductor devices are generally fabricated on GaAs wafers, however, for certain optoelectronic applications InP wafers are also used. The electrical and chemical properties of the wafer surface must be well controlled and therefore the preparation of starting wafers is a crucial portion of IC and device manufacturing. In order to obtain high fabrication yields and good device performance, it is very important that the starting wafers be of reproducibly high quality. For example, the front surface must be smooth and flat on both a macro- and micro-scale, because high-resolution patterns (lithography) are optically formed on the wafer. In principle, cutting a crystal into thin slices and polishing one side until all saw marks are removed and the surface appears smooth and glossy could produce a suitable wafer. However, due in part to the brittleness of Si and GaAs crystals, as well as the increasing requirements of wafer cleanliness and surface defect reduction with ever decreasing device geometries, a very complex series of processing steps are required to produce analytically clean, flat and damage-free wafer surfaces.

The following focuses on the general principles and methods with regard to wafer formation. Detailed formulas, recipes, and specific process parameters are not given as they vary considerably among different wafer producers. In general, techniques for fabrication of Si wafers have become standardized within the semiconductor industry. In contrast, GaAs wafer technology is less standardized, possibly due to either:

- similarity to silicon practices,
- lower production volume of GaAs wafers.

There are two general classes of processes in the methodology of making wafers:

- mechanical,
- chemical.

As both Si and GaAs are brittle materials, the mechanical processes for their wafer fabrication are similar; however, the different chemistry of Si and GaAs require that the chemical processes be dealt with separately.

Each of the processing steps in the conversion of a semiconductor *ingot* (formed by Czochralski or Bridgeman growth) into a polished wafer ready for device fabrication, results in the removal of material from the original ingot; between $^1/_3$ and $^1/_2$ of the original ingot is sacrificed during processing. Methods for the removal of material from a crystal ingot are classified depending on the size of the particles being removed during the process.

If the removed particles are much larger than atomic or molecular dimensions the process is described as being macro-scale. Conversely, if the material is removed atom-by-atom or molecule-by-molecule then the process is termed micro-scale. A further distinction between various types of processes is whether the removal occurs as a result of mechanical or chemical processes. The formation of a finished wafer from a semiconductor ingot normally requires six machining (mechanical) operations, two chemical operations, and at least one polishing (chemical-mechanical) operation. Additionally, multiple inspection and evaluation steps are included in the overall process. A summary of the individual steps, and their functions, involved in wafer production is shown in Table 5.1.

Process	Type	Function
Cropping	Mechanical	Removal of conical shaped ends and impure portions
Grinding	Mechanical	Obtain precise diameter
Orientation flatting	Mechanical	Identification of crystal orientation and dopant type
Etching	Chemical	Removal of surface damage
Wafering	Mechanical	Formation of individual wafers by cutting
Heat treatment	Thermal	Annihilation of undesirable electronic donors
Edge contouring	Mechanical	Provide radius on the edge of the wafer
Lapping	Mechanical	Provides requisite flatness of the wafer
Etching	Chemical	Removal of surface damage
Polishing	Mechano-chemical	Provides a smooth (specular) surface

Cleaning	Chemical	Removal of organics, heavy metals, and particulates

Table 5.1: Summary of the process steps involved in semiconductor wafer production.

Crystal shaping

Although an as-grown crystal ingot is of high purity (99.9999%) and crystallinity, it does not have the sufficiently precise shape required for ready wafer formation. Thus, prior to slicing an ingot into individual wafers, several steps are needed. These operations required to prepare the crystal for slicing are referred to as crystal shaping, and are shown in Figure 5.1.

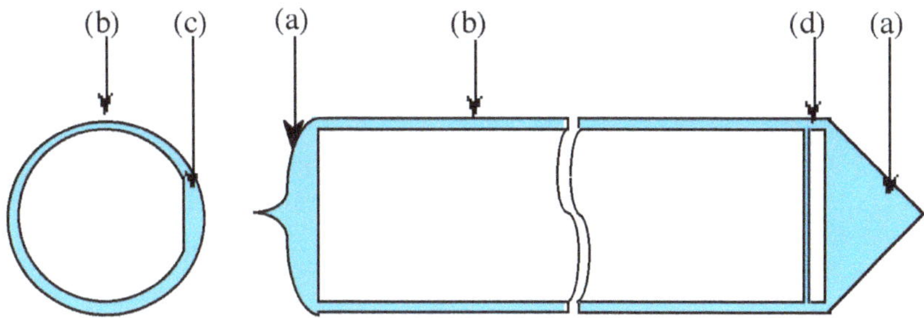

Figure 5.1: Schematic representation of crystal shaping operations: (a) remove crown and taper, (b) grind to required diameter, (c) grind flat, and (d) slice sample for measurements. Shaded area represents material removed.

Cropping

The as-grown ingots have conical shaped *seed* (top) and *tang* (bottom) ends that are removed using a circular diamond saw for ease of further manipulation of the ingot (Figure 5.1a). The cuttings are sufficiently pure that they are cleaned and the recycled in the crystal growth operation. Portions of the ingot that fail to meet specifications of resistivity are also removed. In the case of silicon ingots these sections may be sold as metallurgical-grade silicon (MGS). Conversely, portions of the crystal that meet desired resistivity specifications may be preferentially selected. A sample slice is also cut to enable oxygen and carbon content to be determined; usually this is accomplished by Fourier transform infrared spectroscopic measurements (FT-IR). Finally, cropping is used to cut crystals to a suitable length to fit the saw capacity.

Grinding

The primary purpose of crystal grinding is to obtain wafers of precise diameter because the automatic diameter control systems on crystal growth equipment are not capable of meeting the tight wafer diameter specifications. In addition, crystals are seldom grown perfectly round in cross section. Thus, ingots are usually grown with a 1-2 mm allowance and reduced to the proper diameter by grinding Figure 5.1b.

Crystal grinding is a straightforward process using an abrasive grinding wheel, however, it must be well controlled in order to avoid problems in subsequent operations. Exit chipping in wafering and lattice slip in thermal processing are problems often resulting from improper crystal grinding. Two methods are used for crystal grinding:
- grinding on center,
- centerless grinding.

Figure 5.2 shows a schematic of the general set-up for grinding a crystal ingot on center. The crystal is supported at each end in a lathe-like machine. The rotating cutting tool, employing a water-based coolant, makes multiple passes down the rotating ingot until the requisite diameter is obtained. The center grinder can also be used for grinding the identification flats as well as providing a uniform ingot diameter. However, grinding the crystal on centers requires that the operator locate the crystal axis in order to obtain the best yield.

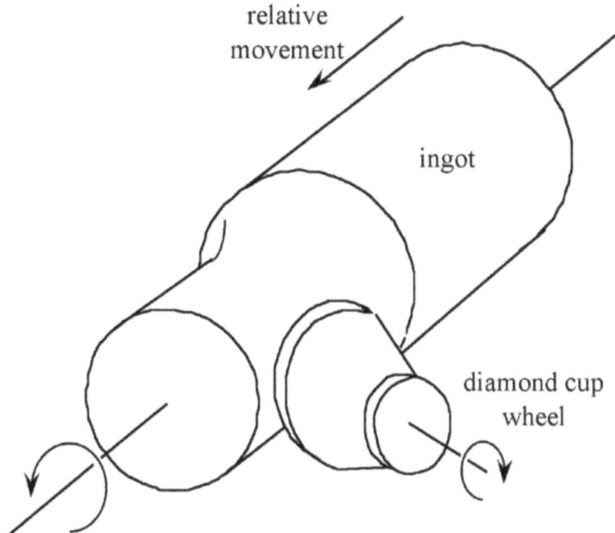

Figure 5.2: Schematic representation of grinding on center.

Centerless grinding eliminates the problems associated with locating the crystal center. The centerless method is superior for long crystals; however, a centerless grinder is much larger than a center grinder of the same diameter capacity. In centerless grinding the ingot is supported between two wheels, a grinding wheel and a drive wheel. A schematic of the centerless grinder is shown in Figure 5.3. The axis of the drive wheel is canted with respect to that of the crystal ingot and the grinding wheel pushing the crystal ingot past the stationary (but rotating) grinding wheel, see Figure 5.3b.

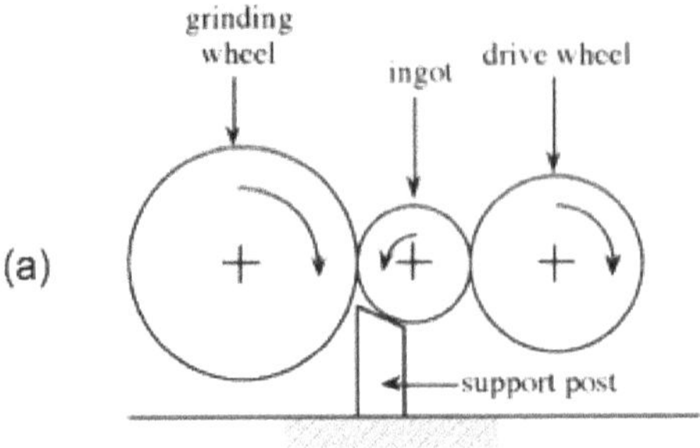

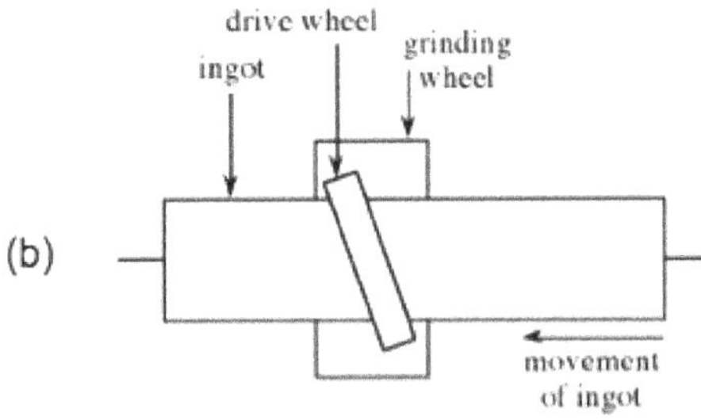

Figure 5.3: Schematic representation of centerless grinding viewed (a) along and (b) perpendicular to the crystal axis.

Orientation/identification flats

Following grinding of the ingot to the desired diameter, one or two flats are ground along the length of the ingot. The identification flats (one or two) are ground lengthwise along the crystal according to the orientation and the dopant type. After grinding the crystal on centers the crystal is rotated to the proper orientation, then the wheel is positioned with its axis of rotation perpendicular to the crystal axis and moved along the crystal from end to end until the appropriate flat size is obtained. An optical or X-ray orientation fixture may be used in conjunction with the crystal mounting to facilitate the proper orientation of the crystal on the grinder.

The largest flat is called the primary flat (Figure 5.1c) and is parallel to one of the crystal planes, as determined by X-ray diffraction. The primary flat is used for automated positioning of the wafer during subsequent processing steps, e.g., lithographic patterning and dicing. Other smaller flats are called "secondary flats" and are used to identify the crystal orientation ($<111>$ versus <100>) and the material (n-type versus p-type). Secondary flats provide a quick and easy manner by which unknown wafers can be sorted. The flats shown schematically in Figure 5.4 are located according to a Semiconductor Equipment and Materials Institute (SEMI®) standard and are ground to specific widths, depending upon crystals diameter. Notches are also used in place of the secondary flat; however, the relative orientations of the notch and primary flat with regard to crystal orientation and dopant are maintained.

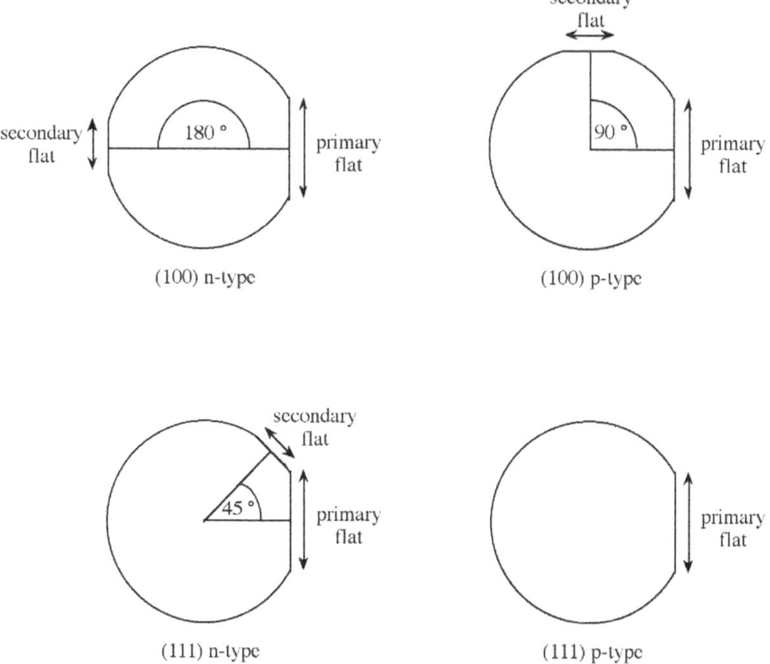

Figure 5.4: SEMI locations for orientation/identification flats.

Etching

The cropping and grinding processes are performed with relatively coarse abrasive and consequently a great deal of subsurface damage results. Pits, chips, and cracks all contribute to stress in the cut wafer and provide nuclei for crack propagation at the edges of the finished wafer. If regions of stress are removed then cracks will no longer propagate, reducing exit chipping and wafer breakage during subsequent fabrication steps.

The general method for removing surface damage is to etch the crystal in a hot solution. The most common etchants for Si are based on the HNO_3-HF system, in which etchant modifiers such as acetic acid also commonly used. In the case of GaAs, HCl-HNO_3 is the appropriate system. These etchants selectively attack the crystal at the damaged regions. After etching, the crystal is transferred to the slicing preparation area.

Wafering

The purpose of *wafering* is to saw the crystal into thin slices with precise geometric dimensions. By far, the most common method of wafering semiconductor crystals is the use of an annular, or inner diameter (ID), diamond saw blade. A schematic diagram of ID slicing technology is shown in Figure 5.5.

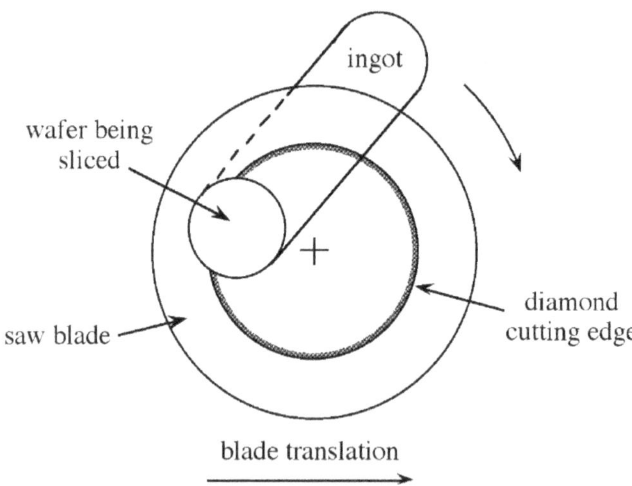

Figure 5.5: Schematic diagrams of ID slicing process.

The crystal, when it arrives at the sawing area, has been ground to diameter, flatted, and etched. In order to slice it, the crystal must be firmly mounted in such a way that it can be completely converted to wafers with minimum waste. The crystal is attached with wax or epoxy to a mounting block, which is usually cylindrical in shape and of the same diameter as the ingot. Also, a mounting beam (or strip) is attached along the length of the crystal at the breakout point of the saw blade. This reduces exit chipping (breakage that occurs as the blade exits the crystal at the end of a cut) and also provides support for the sawn wafer until it is retrieved. Graphite or phenolic resins are common materials for the mounting block and beams, although some success has been obtained in mounting ingots using hydraulic pressure.

The saw blade is a thin sheet of stainless steel (325 µm), with diamond bonded to its inner edge. This blade is mounted on a drum that rotates at ca. 2000 rpm. Saw blades 58 cm ($\approx$23 inches) in diameter with a 20 cm (8 inches) opening are common, however, as wafer sizes increase larger blades are employed: 30 cm (12 inches) wafers are common for Si. The blade moves relative to the stationary crystal at a speed of 0.05 cm/s, and the cutting process is water-cooled. Thus, considering that wafers are sliced sequentially (one at a time), the overall process is very slow.

A further problem is that the kerf loss (loss due to the width of the blade) results in approximately $^1/_3$ of the material being lost as saw dust. Finally, the depth of the drum onto which the blade is attached limits the length of the ingot section that is accessible. In order to overcome this problem, another style of ID blade saw was developed in which the blade is mounted on an air bearing and is rotated by a belt drive. This allows the entire length of the crystal ingot to be sliced.

Both silicon and GaAs crystals are grown with either the crystallographic <100> or <111> direction parallel to the cylindrical axis of the crystal. Wafers may be cut either exactly perpendicular to the crystallographic axis or deliberately off-axis by several degrees. In order to obtain the proper wafer orientation, the crystal must be properly oriented on the saw. All production slicing-machines have adjustments for orientation of the crystal; however, it is usually necessary to check the orientation of the first slice in order to assure that all subsequent slices will be properly oriented.

Obvious variables introduced during the wafering process include: cutting rate, wheel speed, and coolant flow rate; however, the condition of the machines, such as alignment and vibration, is the most important variable followed by the condition of the blade. A deviated blade rim may cause taper, bow, or warp. Table 5.2 summarizes the types of deformations that can occur during wafering, their physical appearance and their characteristics.

Type of bow and warp	Surface appearance	Lattice curvature	Comments

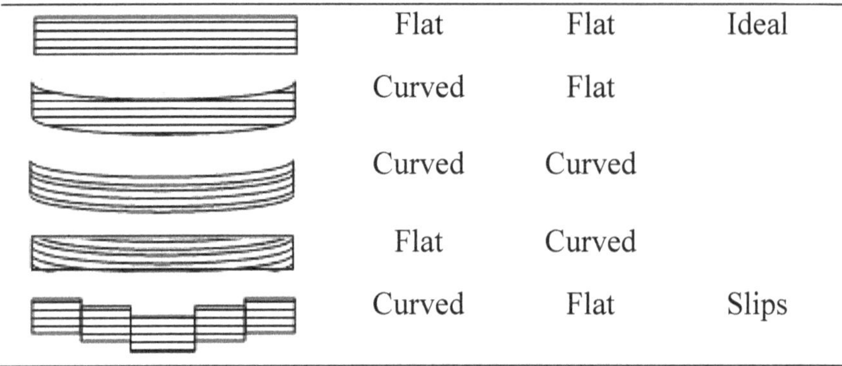

Figure 5.2: Deformed wafers and their characteristics.

Heat treatment

As-produced Czochralski grown crystals often have a level of oxygen impurity that may exceed the concentration of dopant in the semiconductor material (i.e., Si or GaAs). This oxygen impurity has a deleterious effect on the semiconductor properties, especially upon subsequent thermal processing, e.g., thermal oxide growth or epitaxial film growth by metal organic chemical vapor deposition (MOCVD). For example, when silicon crystals are heated to about 450 °C the oxygen undergoes a transformation that causes it to behave as an electron donor, much like an n-type dopant. These oxygen donors, or *thermal donors*, mask the true resistivity of the semiconductor because they either add additional carrier electrons to a n-type crystal or compensate for the positive holes in a p-type crystal. Fortunately, these thermal donors can be *annihilated* by heat-treating the materials briefly in the range of 500 - 800 °C and then cooling quickly through the 450 °C region before donors can reform.

In principle thermal donor annihilation can be performed on wafers at any time during their fabrication; however, it is usually best to perform the heat treatment immediately after wafering since sub-standard wafers may be rejected before additional processing steps are undertaken and thus limiting additional cost. Donor annihilation is a bulk effect, and therefore the thermal treatment can be performed in air, since any surface oxide that may form will be removed in subsequent lapping and polishing steps.

Lapping or grinding

The as-cut wafers vary sufficiently in thickness to require an additional operation, the slicing operation does not consistently produce the required flatness and parallelism required for many wafer specifications, see Table 5.2. Since conventional polishing does not correct variations in flatness or thickness, a mechanical two-sided lapping operation is performed. Lapping is capable of achieving very precise thickness uniformity, flatness and parallelism. Lapping also prepares the surface for polishing by removing the sub-surface sawing damage, replacing it with a more uniform and smaller lapping damage.

The process used for lapping semiconductor wafers evolved from the optical lens manufacturing industry, using principles developed over several hundred years; however, as the lens has a curved surface and the wafers are flat, the equipment for lapping wafers is mechanically simpler than lens processing machines. The simplest double-side lapping machine consists of two very flat counter-rotating plates, carriers to hold and move the wafers between the plates, and a device to feed abrasive slurry steadily between the plates. The abrasive is typically 9 μm Al_2O_3 grit. Commercial abrasives are suspended in water or glycerin with proprietary additives to assist in suspension and dispersion of the particles, to improve the flow properties of the slurry, and to prevent corrosion of the lapping machine. Hydraulics or an air cylinder applies lapping pressure with low starting pressure for 2-5 minutes, which is then increased through most of the process. The completion of lapping may be determined by elapsed time or by an external thickness-sensing device. The finished process gives a wafer with a surface uniform to within 2 μm. Approximately 20 μm per side is removed during the lapping process.

Although lapping would appear to be simple in concept, the successful implementation of a production lapping operation requires the development of a technique and experience to achieve acceptable quality with good yields. Small adjustments to the rotation rates of the plates and carriers will cause the plates to wear concave, convex or flat.

As lapping is a messy process, various efforts have been made to avoid it or to substitute an alternative process. The most likely approach at present is grinding, in which the wafer is held on a vacuum chuck and a series of progressively finer diamond wheels is moved over the wafer while it is rotated on a turntable. Grinding gives a clearer surface than lapping, however, only one side may be ground at a time and the resulting flatness is not as good as that obtained by lapping.

Edge contouring

The rounding of the edge of the wafer to a specific contour is a fairly recent development in the technology of wafer preparation. It was known by the early seventies that a significant number of device yield problems could be traced to the physical condition of the wafer edge. An acute edge affects the strength of the wafer due to: stress concentration, and a lowering of its resistance to thermal stress, as well as being the source of particle chip, breakage, and lattice damage. In addition, the particles originating from the chipped edges can, if present on the wafer surface, add to the defect density (D_0) of the IC process reducing fabrication yield. Further problems associated with a square edge include the build-up of photoresist at the wafer edge. The solution to these process problems is to provide a contoured edge with a defined radius (r).

Chemical etching of wafers results in a degree of edge rounding, but it is difficult to control. Thus, mechanical edge contouring has been developed and the result has been a dramatic improvement in yields in downstream wafer processing. Losses due to wafer breakage are also reduced. The edge contouring process is usually performed in cassette-fed high-speed equipment, in which each wafer is rotated rapidly against a shaped cutting tool (Figure 5.6).

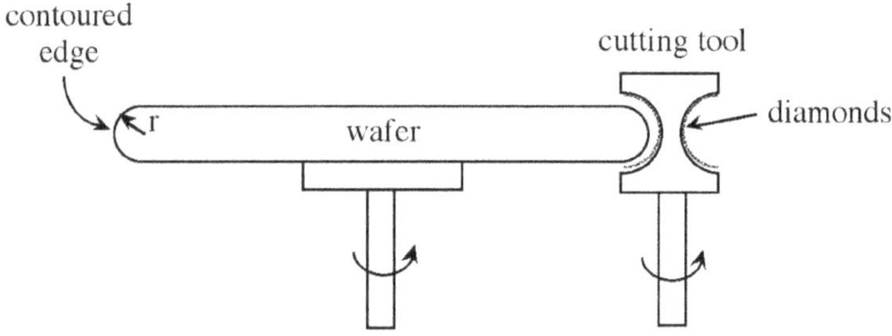

Figure 5.6: Schematic illustration of edge contouring.

Etching

The mechanical processes described above to shape the wafer leave the surface and edges damaged and contaminated. The depth of the work damage depends on the specific process, however, 10 µm is typical. Such damage is readily removed by chemical etching. Etching is used at multiple points during the fabrication of a semiconductor device. The discussion below is limited to etches suitable for wafer fabrication, i.e., non-selective etching of the entire wafer surface.

Wet chemical etching

The wet chemical etching of any material can be considered to involve three steps:

(1) transportation of the reactants to the surface,
(2) reaction at the surface,
(3) movement of the reaction products into the etchant solution (Figure 5.7).

Each of these may be the rate limiting step and thus control the etch rate and uniformity. This effect is summarized in Table 5.3.

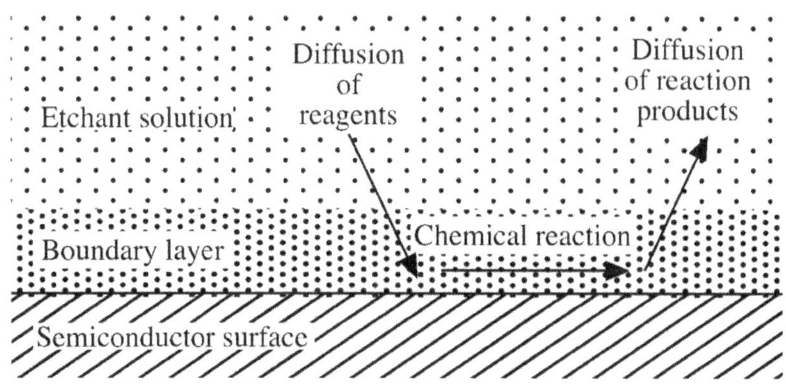

Figure 5.7: Schematic representation of the three steps involved in wet chemical etching: (i) diffusion of the chemical etch reagents through the boundary layer, (ii) chemical reaction at the surface, and (iii) diffusion of the reaction products into the etch solution through the boundary layer.

Rate limiting step	Etching rate	Results	Comments
diffusion of reagent to	slow	etching	enhanced surface

the surface reaction at semicon-ductor surface	fast	(anisotropic) polishing (isotropic)	roughness ideal
diffusion of reaction products from the sur-face	slow	polishing (isotropic)	reaction product remains on sur-face

Table 5.3: Effects of rate limiting step in semiconductor etching.

An etchant that is limited by the rate of reaction at the surface will tend to enhance any surface features and promote surface roughness due to preferential etching at defects (anisotropic). In contrast, if the etch rate is limited by the diffusion of the etchant reagent through a stagnant (dead) boundary layer near the surface, then the etch will result in uniform polishing and the surface will become smooth (isotropic). If removal of the reaction products is rate limiting then the etch rate will be slow because the etch equilibrium will be shifted towards the reactants. In the case of an individual etchant reaction, the rate determining step may be changed by rapid stirring to aid removal of reaction products, or by increasing the temperature of the etch solution, see Figure 5.8. The exact etching conditions are chosen depending on the application. For example, dilute high temperature etches are often employed where the etch damage must be minimized, while cooled etches can be used where precise etch control is required.

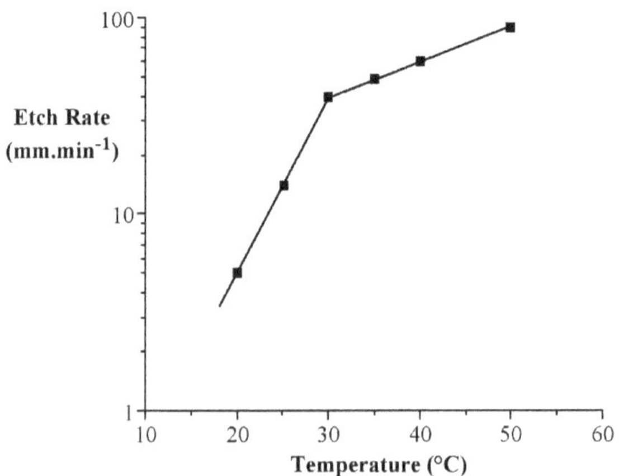

Figure 5.8: Typical etch rate versus temperature plot for a mixture of HF (20%), nitric acid (45%), and acetic acid (35%).

Traditionally mixtures of hydrofluoric acid (HF), nitric acid (HNO_3) and acetic acid ($MeCO_2H$) have been used for silicon, but alkaline etches using potassium hydroxide (KOH) or sodium hydroxide (NaOH) solutions are increasingly common. Similarly, gallium arsenide etches may be either acidic or basic; however, in both cases, the etches are oxidative due to the use of hydrogen peroxide.

A wide range of chemical reagents is commercially available in "transistor grade" purity and these are employed to minimize contamination of the semiconductor. Deionized water is commonly used as a diluent for each of these reagents and the concentration of commonly used aqueous reagents is given in Table 5.4.

Reagent	Weight %
HCl	37
H_2SO_4	98
HNO_3	79
$MeCO_2H$	99
NH_4OH	29
HF	49
H_3PO_4	85
$HClO_4$	70
H_2O_2	30

Table 5.4: Weight percent concentration of commonly used concentrated aqueous reagents.

The equipment used for a typical etchant process includes an acid (or alkaline) resistant tank, which contains the etchant solution and one or more positions for rinsing the wafers with deionized water. The process is batch in nature involving tens of wafers and the best equipment provides a means of rotating the wafers during the etch step to maintain uniformity. In order to assure the removal of all surface damage, substantial over-etching is performed. Thus, the removal of 20 µm from each side of the wafer is typical. Etch times are usually several minutes per batch.

Etching silicon

The most commonly used etchants for silicon are mixtures of hydrofluoric acid (HF) and nitric acid (HNO_3) in water or acetic acid ($MeCO_2H$). The etching involves a reduction-oxidation (redox) reaction, followed by dissolution of the reaction products. In the HF-HNO_3 system the HNO_3 oxidizes the silicon and the HF removes the reaction products from the surface. The overall reaction is:

$$Si + HNO_3 + 6\,HF \rightarrow H_2SiF_6 + HNO_2 + H_2O$$

The oxidation reaction involves the oxidation of Si^0 to Si^{4+}, and it is auto-catalytic in that the reaction product promotes the reaction itself. The initial step involves trace impurities of HNO_2 in the HNO_3 solution,

$$HNO_2 + HNO_3 \rightarrow N_2O_4 + H_2O$$

which react to liberate nitrogen dioxide (NO_2),

$$N_2O_4 \rightarrow 2\,NO_2$$

The nitrogen dioxide oxidizes the silicon surface in the presence of water, resulting in the formation of $Si(OH)_2$ and the reformation of HNO_2,

$$Si^0 + 2\,NO_2 + 2\,H_2O \rightarrow Si(OH)_2 + 2\,HNO_2$$

The $Si(OH)_2$ decomposes to give SiO_2,

$$Si(OH)_2 \rightarrow SiO_2 + H_2$$

Since the reaction between HNO_2 and HNO_3 is rate limiting, an induction period is observed. However, this is overcome by the addition of NO_2^- ions in the form of $[NH_4][NO_2]$.

The final step of the etch process is the dissolution of the SiO_2 by HF,

$$SiO_2 + 6\,HF \rightarrow H_2SiF_6 + H_2O$$

Stirring serves to remove the soluble products from the reaction surface. The role of the HF is to act as a *complexing reagent*, and thus the reaction is known as a *complexing reaction*. The formation of water as a reaction product requires that acetic acid be used as a diluent (solvent) to ensure better control.

The etching reaction is highly dependent on the relative ratios of the etchant reagents. Thus, if an HF-rich solution is used, the reaction is limited by the oxidation step and the etching is anisotropic, since the oxidation reaction is sensitive to doping, crystal orientation, and defects. In contrast, the use of an HNO_3-rich solution produces isotropic etching since the dissolution process is rate limiting (Table 5.3). The reaction of HNO_3-rich solutions has been found to be diffusion-controlled over the temperature range 20-50 °C (Figure 5.7), and is therefore commonly employed for removing work damage produced during wafer fabrication. The boundary layer thickness (Figure 5.7), and therefore the dimensional control over the wafer, is regulated by the rotation rate of the wafers. A common etch formulation is a 4:1:3 mixture of HNO_3 (79%), HF (49%), and acetic acid ($MeCO_2H$, 99%). There are some etchant formulations that are based on alternative (or additional) oxidizing agents, such as: Br_2, I_2, and $KMnO_4$.

Alkaline etching (KOH/H_2O or $NaOH/H_2O$) is by nature anisotropic and the etch rate depends on the number of dangling bonds which in turn are dependent on the surface orientation. Since etching is reaction rate limited no rotation of the wafers is necessary and excellent uniformity over large wafers is obtained. Alkaline etchants are used with large wafers where dimensional uniformity is not maintained during lapping. A typical formulation uses KOH in a 45% weight solution in H_2O at 90 °C.

Etching gallium arsenide

Although a wide range of etches have been investigated for GaAs, few are truly isotropic, because the surface activity of the (111) Ga and (111) As face are very different. The As rich face is considerably more reactive than the Ga rich face, thus under identical conditions it will etch faster. As a result most etches give a polished surface on the As face, but the Ga face tends to appear cloudy or frosted due to the highlighting of surface features and crystallographic defects.

As with silicon the etch systems involve oxidation and complexation; however, in the case of GaAs the gallium is already fully oxidized (formally Ga^{3+}), thus, it is the arsenic (formally the arsenide ion, As^{3-} that is oxidized by a suitable oxidizing agent (e.g., H_2O_2) to the soluble oxide, As_2O_3,

$$2\,As^{3-} + 6\,H_2O_2 + H^+ \rightarrow As_2O_3 + 5\,OH^- + 4\,H_2O$$

The gallium ions form the oxide Ga_2O_3 via the hydroxide,

$$2\,Ga^{3+} + 6\,OH^- \rightarrow 2\,Ga(OH)_3 + 3\,H_2O$$

Both oxides are soluble in acid solutions, resulting in their removal from the surface.

The peroxide based oxidative etches for GaAs are divided into acidic and basic etches. The composition and application of some of these systems are summarized in Table 5.5. The most widely used of these is $H_2SO_4/H_2O_2/H_2O$ and is referred to as Caro's acid. The high viscosity of H_2SO_4 results in diffusion-limited etching with high acid concentrations. Etches with low acid concentrations tend to be anisotropic. Phosphoric acid (H_3PO_4) or citric acid (Figure 5.9) may be exchanged for sulfuric acid (H_2SO_4). Replacement of the acid component with bases such as NH_4OH or $NaOH$ can result in near to truly isotropic etchants, although certain combinations can result in strong anisotropy.

Formulation	Volume ratio	(100) etch rate (μm/min)	(110) etch rate (μm/min)	(111) As etch rate (μm/min)	(111) Ga etch rate (μm/min)
$H_2SO_4/$ $H_2O_2/$ H_2O	8:1:1	1.5	1.5	1.5	0.8
$H_2SO_4/$ $H_2O_2/$ H_2O	1:8:1	8.0	8.0	12.0	3.0
$H_2SO_4/$ $H_2O_2/$ H_2O	3:1:50	0.8	0.8	0.8	0.4
citric acid/ $H_2O_2/$ H_2O	1:1:1	0.6	0.6	0.6	0.4

| NH$_4$OH/ H$_2$O$_2$/ H$_2$O | 1:700 | 0.3 | 0.3 | 0.3 | 0.3 |
| NaOH/ H$_2$O$_2$/ H$_2$O | 1:0.76 | 0.2 | 0.2 | 0.2 | 0.2 |

Table 5.5: The composition and application of selected etch systems for GaAs.

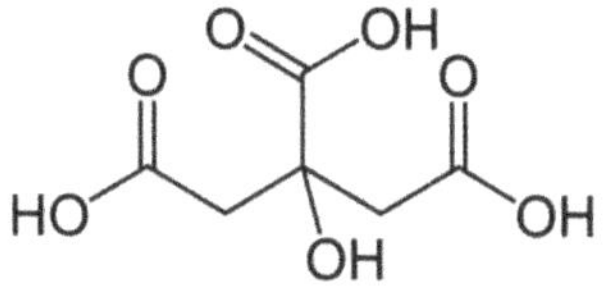

Figure 5.9: Structure of citric acid.

One of the earliest etching systems for GaAs is based on the use of a dilute (ca. 0.05 vol.%) solution of bromine (Br$_2$) in ethanol. The Br$_2$ acts as the oxidant, resulting in the formation of soluble bromides. The etch rate of this system is different for different crystallographic planes, i.e., the etch rates for the (111) As, (100), and (111) Ga faces are in the ratio 6:5:1, although more uniform etch rates are observed with high Br$_2$ concentrations (ca. 10 vol.%). These higher concentration solutions are used for the removal of damage due to cutting with the saw.

Polishing

The purpose of polishing is to produce a smooth, specular surface on which device features can be defined by lithography. In order to allow for very large scale integration (VLSI) or ultra large scale integration (ULSI) fabrication the wafer must have a surface with a high degree of flatness. Variations less than 5-10 μm across the wafer diameter are typical flatness specifications. In addition, given the preceding steps, wafer polishing must not leave residual contamination or surface damage. The techniques of wafer polishing are derived from the glass lens industry, with some important modifications that have been developed to meet the special requirements of the microelectronics industry.

Differences between polishing and lapping

If the surface of a wafer that has undergone lapping (or grinding) is examined with an electron microscope, cracks, ridges and valleys are observed. The top "relief layer" consists of peaks and valleys. Below this layer is a damaged layer characterized by microcracks, dislocations, slip and stress. Figure 5.10 shows a schematic representation of the abraded surface. Both of these layers must be removed completely prior to further fabrication.

Decreasing the particle size of the abrasive during lapping only decreases the scale of the damage, but does not eliminate it entirely. In fact this surface damage is a characteristic of the brittle fracture of single crystal Si and GaAs, and occurs because during lapping the abrasive grains are moved across the surface under a pressure beyond that of the fracture strength of the wafer materials (Si or GaAs). In contrast to the mechanical abrasion employed in lapping, polishing is a mechano-chemical process during which brittle fracture does not occur. A polished wafer does not display any evidence of a relief surface such as that produced by lapping, even at highest resolution electron microscope.

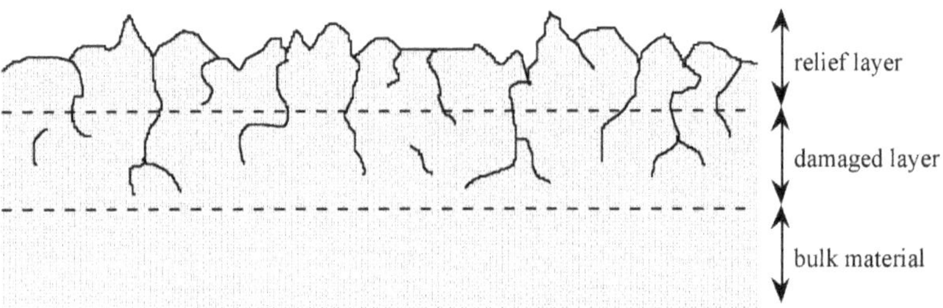

Figure 5.10: Schematic representation of a cross sectional view of an abraded wafer surface prior to polishing.

Process of Polishing

Figure 5.11 shows a schematic of the polishing process. Polishing may be conducted on single wafers or as a batch process depending on the equipment employed. Single wafer polishing is preferred for larger wafers and allows for better surface flatness. In both processes, wafers are mounted onto a fixture, by either wax or a composite Felx-Mount™, and pressed against the polishing pad. The polishing pad is usually made from an artificial fabric such as polyester felt-polyurethane laminate.

Polishing is accomplished by a mechano-chemical process, in which aqueous polishing slurry is dripped onto the polishing pad (Figure 5.11). The polishing slurry performs both a chemical and mechanical process, and consists of fine silica (SiO_2) particles (100 Å diameter) and an oxidizing agent. Aqueous sodium hydroxide (NaOH) is used for Si, while aqueous sodium chlorate (NaOCl) is preferred for GaAs. Suspending agents are usually added to prevent settling of the silica particles. Under the heat caused by the friction of the wafer on the polishing pad the wafer surface is oxidized, which is the chemical step, while in the mechanical step the silica particles in the slurry abrade the oxidized surface away.

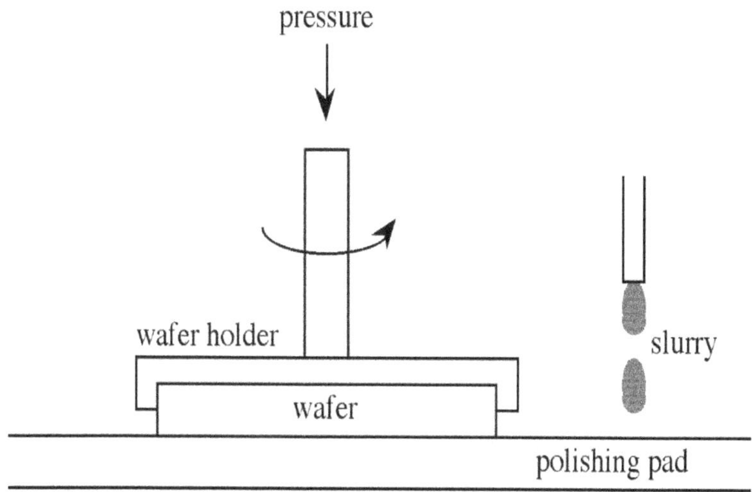

Figure 5.11: Schematic representation of the wafer polishing process.

In order to achieve a reasonable rate of removal of the relief and damaged layers and still obtain the highest quality surface, the polishing is done in two steps, stock removal and haze removal. The former is carried out with a higher concentration slurry and may proceed for about 30 minutes at a removal rate of 1 µm/min. Haze removal is performed with a very dilute slurry, a softer pad with a reaction time of about 5-10 minutes, during which the total amount of material removed is only about 1 µm. Due to the active chemical reaction between the wafer and the polishing agent, the wafers must be rinsed in deionized water immediately after polishing to prevent haze or stains from reforming.

There are many variables that will influence the rate and quality of polishing. High pressure results in a higher polishing rate, but excessive pressure may cause non-uniform polishing, excessive heat generation and fast pad wear. The rate of polishing is increased with higher temperatures but this may also lead to haze formation. High wheel speeds accelerate the polishing rate but can raise the temperature and also results in problems in maintaining a uniform flow of slurry across the pad. Dense slurry concentrations increase the polishing rate but are more costly. The pH of the slurry solution can also affect the polishing rate, for example the polishing rate of Si gradually increases with increased pH (higher basicity) until a pH of about 12 where a dramatic decrease is observed. In general, the optimum polishing process for a given facility depends largely upon the interplay of product specification, yields, cost, and quality considerations and must be developed uniquely. The wafer polishing process does not improve the wafer flatness and, at best, polishing will not degrade the wafer flatness achieved in the lapping operation.

Cleaning

During the processes described above, semiconductor wafers are subjected to physical handling that leads to significant contamination. Possible sources of physical contamination include:

- airborne bacteria,
- grease and wax from cutting oils and physical handling,
- abrasive particulates (usually, silica, silicon carbide, alumina, or diamond dust) from lapping, grinding or sawing operations,
- plasticizers which are derived from containers and wrapping in which the wafers are handled and shipped.

Chemical contamination may also occur as due to improper cleaning after etch steps. Light-metal (especially sodium and potassium) species may be traced to impurities in etchant solutions and are chemisorbed on to the surface where they are particularly problematical for metal oxide semiconductor (MOS) based devices, although higher levels of such impurities are tolerable for bipolar devices. Heavy metal impurities (e.g., Cu, Au, Fe, and Ag) are usually caused by electrodeposition from etchant solutions during fabrication. While wafers are cleaned prior to shipping, contamination accumulated during shipping and storage necessitates that all wafers be subjected to scrupulous cleaning prior to fabrication. Furthermore, cleaning is required at each step during the fabrication process. Although wafer cleaning is a vital part of each fabrication step, it is convenient to discuss cleaning within the general topic of wafer fabrication.

Cleaning silicon

The first step in cleaning a Si wafer is removal of all physical contaminants. These contaminates are removed by rinsing the wafer in hot organic solvents such as 1,1,1-trichloroethane (Cl_3CCH_3) or xylene ($C_6H_4Me_2$), accompanied by mechanical scrubbing, ultrasonic agitation, or compressed gas jets. Removal of the majority of light metal contaminants is accomplished by rinsing in hot deionized water; however, complete removal requires a further more aggressive cleaning process. The most widely used cleaning method in the Si semiconductor industry is based on a two-step, two-solution sequence known as the "RCA Cleaning Method".

The first solution consists of H_2O-H_2O_2-NH_4OH in a volume ratio of 5:1:1 to 7:2:1, which is used to remove organic contaminants and heavy metals. The oxidation of the remaining organic contaminants by the hydrogen peroxide (H_2O_2) produces water-soluble products. Similarly, metal contaminants such as cadmium, cobalt, copper, mercury, nickel, and silver are solubilized by the NH_4OH through the formation of soluble amino complexes, e.g.,

$$2\,Cu^{2+}\,(s)\ +6\,NH_4OH\,(aq) \rightarrow [Cu(NH_3)_6\,(aq)$$

The second solution consists of H_2O-H_2O_2-HCl in a 6:1:1 to 8:2:1 volume ratio and removes the Group I (1), II (2) and III (13) metals. In addition, the second solution prevents re-deposition of the metal contaminants. Each of the washing steps is carried out for 10-20 min. at 75-85 °C with rapid agitation. Finally, the wafers are blown dry under a stream of nitrogen gas.

Cleaning GaAs

In principle GaAs wafers may be cleaned in a similar manner to silicon wafers. The first step involves successive cleaning with hot organic solvents such as 1,1,1-trichloroethane, acetone, and methanol, each for 5-10 minutes. GaAs wafers cleaned in this manner may be stored under methanol for short periods of time.

Most cleaning solutions for GaAs are actually etches. A typical solution is similar to the second RCA solution and consists of an 80:10:1 ratio of H_2O-H_2O_2-HCl. This solution is generally used at elevated temperatures (70 °C) with short dip times since it has a very fast etch rate (4.0 μm/min).

Measurements, inspections and packaging

Quality control measurements of the semiconductor crystal and subsequent wafer are performed throughout the process as an essential part of the fabrication of wafers. From crystal and wafer shaping through the final wafer finishing steps, quality control measurements are used to ensure that the materials meets customer specifications, and that problems can be corrected before they create scrap material and thus avoid further processing of reject material. Quality control measurements can be broadly classified into mechanical, electrical, structural, and chemical.

Mechanical measurements are concerned with the physical dimensions of the wafer, including: thickness, flatness, bow, taper and edge contour. Electrical measurements usually include: resistivity and lateral resistivity gradient, carrier type and lifetime. Measurements giving information on the perfection of the semiconductor crystal lattice are classified in the structural category and include: testing for stacking faults, and dislocations. Routine chemical measurements are limited to the measurement of dissolved oxygen and carbon by Fourier transform infrared spectroscopy (FT-IR). Finished wafers are individually marked for the purpose of identification and traceability. Packaging helps protect the finished wafers from contamination during shipping and storage.

Industry standards defining in detail how quality control measurements are to be made, and determining the acceptable ranges for measured values, have been developed by the American Society of Testing Materials (ASTM) and the Semiconductor Equipment and Materials Institute (SEMI).

Chapter 6: Device Fundamentals

Bipolar transistor

The structure of a *bipolar transistor* is shown in Figure 6.1. The device consists of three layers of silicon, a heavily doped n-type layer called the *emitter*, a moderately doped p-type layer called the *base*, and third, more lightly doped layer called the *collector*.

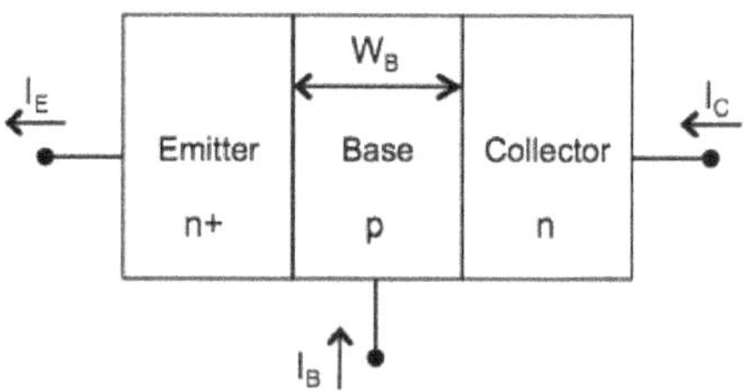

Figure 6.1: Structure of a n-p-n bipolar transistor.

In a biasing (applied DC potential) configuration called *forward active biasing*, the emitter-base junction is forward biased, and the base-collector junction is reverse biased. Figure 6.2 shows the biasing conventions. Both bias voltages are referenced to the base terminal. Since the base-emitter junction is forward biased, and the base is made of p-type material, V_{EB} must be negative. On the other hand, in order to reverse bias the base-collector junction V_{CB} will be a positive voltage.

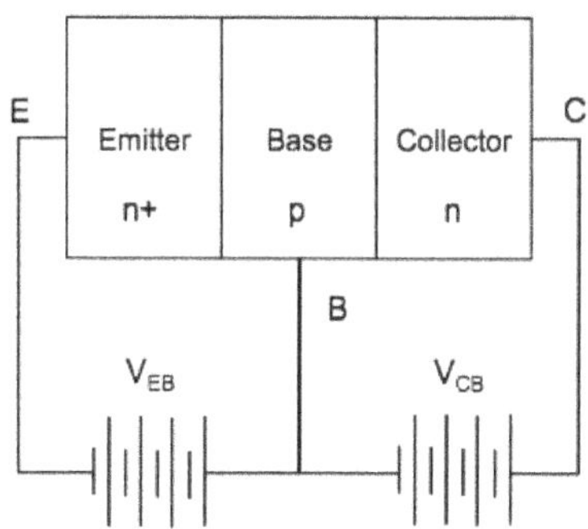

Figure 6.2: Forward active biasing of a n-p-n bipolar transistor.

The band-diagram for the bipolar transistor is shown in Figure 6.3, and shows all of the important features in the operation the transistor. Since the base-emitter junction is forward biased, electrons will go from the (n-type) emitter into the base. Likewise, some holes from the base will be injected into the emitter.

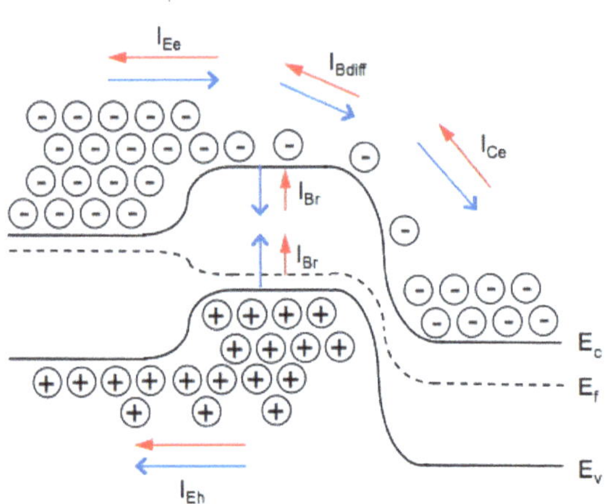

Figure 6.3: Band diagram and carrier fluxes in a bipolar transistor. I_{xy} indicates current flow, where E = emitter, B = base, C = collector, e = electron, h = hole, r = recombination.

In Figure 6.3 there are two different kinds of arrows. The blue open arrows, which are attached to the carriers, show us which way the carrier is moving, while the solid red arrows represent current flow. For holes, motion and current flow are in the same direction, while for electrons, carrier motion and current flow are in opposite directions.

The electrons, which are injected into the base, diffuse away from the emitter-base junction (I_{Ee}) towards the (reverse biased) base-collector junction (I_{Ce}). As they move through the base, some of the electrons encounter holes and recombine with them (denoted by I_{Br}, i.e., current for base recombination). Those electrons which do get to the base-collector junction run into a large electric field which sweeps them out of the base and into the collector (I_{Ce}). They 'fall' down the large potential drop at the junction.

These effects are all seen in Figure 6.3, with arrows representing the various currents, which are associated with each of the carrier's fluxes. I_{Ee} represents the current associated with the electron injection into the base, i.e., it points in the opposite direction from the motion of the electrons, since electrons have a negative charge. I_{Eh} represents the current associated with holes injection into the emitter from the base. I_{Br} represents recombination current in the base, while I_{Ce} represents the electron current going into the collector, such that:

$$I_E = I_{Ee} + I_{Eh}$$

$$I_B = I_{Eh} + I_{Br}$$

$$I_C = I_{Ce}$$

In a 'good' transistor, almost all of the current across the base-emitter junction consists of electrons being injected into the base. Engineers work hard to design the device so that very little emitter current is made up of holes coming from the base into the emitter. The transistor is also designed so that almost all of those electrons, which are injected into the base, make it across to the base-collector reverse-biased junction. Some recombination is unavoidable, but things are arranged so as to minimize this effect.

Basic metal-oxide-semiconductor (MOS) structure

Figure 6.4 shows the basic steps necessary to make the *metal-oxide-semiconductor* (MOS) structure. In order to understand the function of the MOS device it helps if the structure is rotated by 90°, see Figure 6.5. Note that in the p-silicon we have positively charged mobile holes, and negatively charged, fixed acceptors. The band diagram for the semiconductor (Figure 6.5a) is below the schematic of the device (Figure 6.5a). Note that since in this example, the substrate is p-type, the Fermi level is located down close to the valance band. Δ

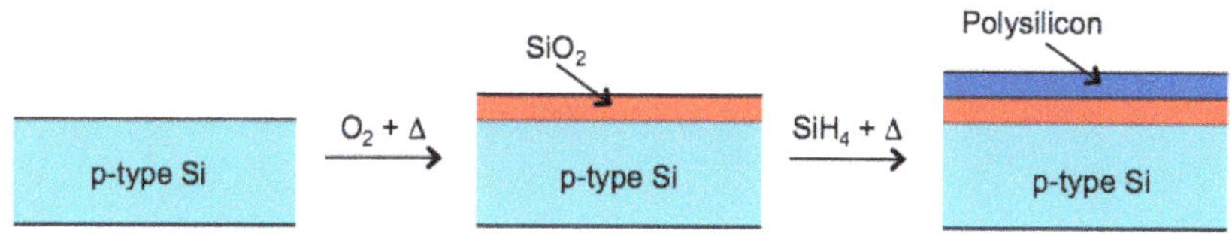

Figure 6.4: Schematic representation of the stepwise formation of the metal-oxide-semiconductor (MOS) structure. Layers not to scale.

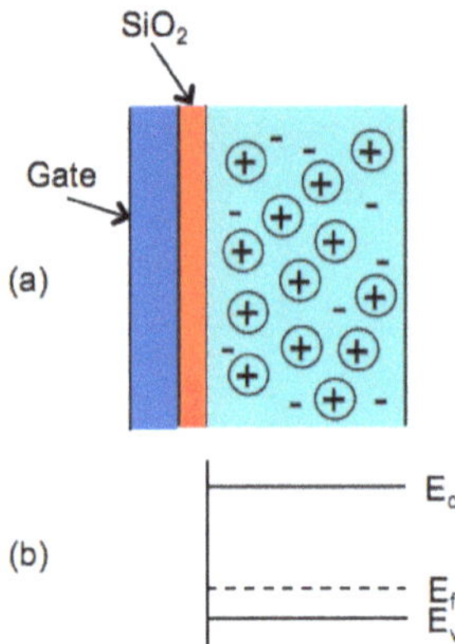

Figure 6.5: Schematic representation of (a) a basic metal-oxide-semiconductor (MOS) structure, and (b) associated band structure of the p-type silicon.

Let us now place a potential between the gate and the silicon substrate. Suppose we make the gate negative with respect to the substrate. Since the substrate is p-type, it has a lot of mobile, positively charged holes in it. Some of them will be attracted to the negative charge on the gate, and move over to the surface of the substrate. This is also reflected in the band diagram shown in Figure 6.6.

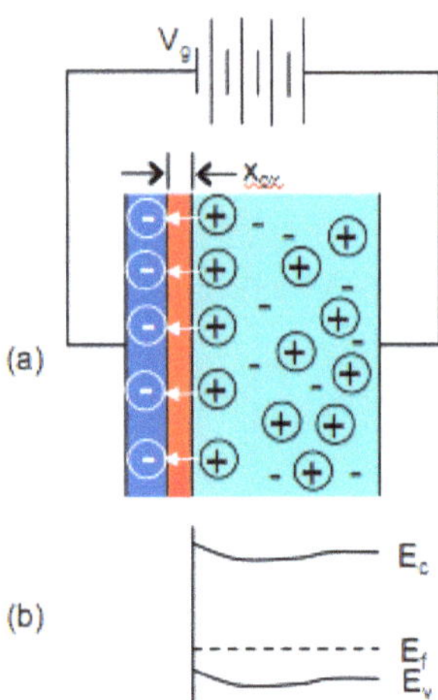

Figure 6.6: Schematic representation of (a) a basic metal-oxide-semiconductor (MOS) structure with applied negative gate voltage, and (b) associated band structure of the p-type silicon layer.

The density of holes is exponentially proportional to how close the Fermi level is to the valence band edge. We see that the band diagram has been 'bent' slightly near the surface to reflect the extra holes, which have accumulated there.

An electric field will develop between the positive holes and the negative gate charge. Note that the gate and the substrate form a parallel plate capacitor, with the oxide acting as the insulating layer in-between them. The oxide is quite thin compared to the area of the device, and so it is quite appropriate to assume that the electric field inside the oxide is a uniform one. The integral of the electric field is just the applied gate voltage V_g. If the oxide has a thickness x_{ox} then since E_{ox} *is* uniform, it is given by,

$$E_{ox} = \frac{V_g}{x_{ox}}$$

If we focus in on a small part of the gate, we can make a cylindrical 'pillbox', which extends from somewhere in the oxide, across the oxide/gate interface and ends up inside the gate material. The pillbox will have an area Δs. Invoking Gauss' law, the surface integral over a closed surface of the displacement vector D (which is $\varepsilon \times E$) is equal to the total charge enclosed by that surface. We will assume that there is a surface charge density $-Q_g$ C/cm^2 on the surface of the gate electrode (Figure 6.7). The integral form of Gauss' Law is:

$$\oint \varepsilon_{ox} E \, dS = Q_{encl}$$

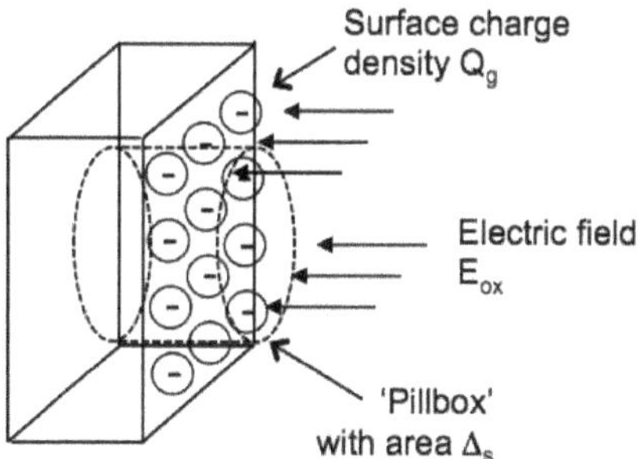

Figure 6.7: Schematic representation of the method of finding the surface charge density.

The charge enclosed in the pillbox is $-(Qg\Delta s)$, and so we have (keeping in mind that the surface integral of a vector pointing into the surface is negative),

$$\oint \varepsilon_{ox} E \, dS = -(\varepsilon_{ox} E_{ox} \Delta(s))$$
$$= -(Q_g \Delta(s))$$

or

$$\varepsilon_{ox} E_{ox} = Q_g$$

Now, we can get,

$$\frac{\varepsilon_{ox} V_g}{x_{ox}} = Q_g$$

or

$$\frac{Q_g}{V_g} = \frac{\varepsilon_{ox}}{x_{ox}} \equiv c_{ox}$$

The quantity c_{ox} is the oxide capacitance, but since it has units of F/cm^2 it is really a capacitance per unit area of the oxide. The dielectric constant of silicon dioxide, ε_{ox}, is about 3.3×10^{-13} F/cm. A typical oxide thickness might be 250 Å (or 2.5×10^{-6} cm). In this case, c_{ox} would be about 1.30×10^{-7} F/cm^2.

The charge on the gate in terms of the gate potential can be found with:

$$Q_g = c_{ox} V_g$$

Although by applying a negative voltage to the gate results in the formation of an *accumulation layer*, this does not achieve the goal of creating a layer of electrons in the MOSFET, which can electrically connect the two n-regions.

Making V_g positive puts a positive charge (Q_g) on the gate, which pushes the holes away from the region under the gate and uncovers some of the negatively charged fixed acceptors. Compared to the case in Figure 6.8 the electric field points the other way, and goes from the positive gate charge, terminating on the negative acceptor charge within the silicon.

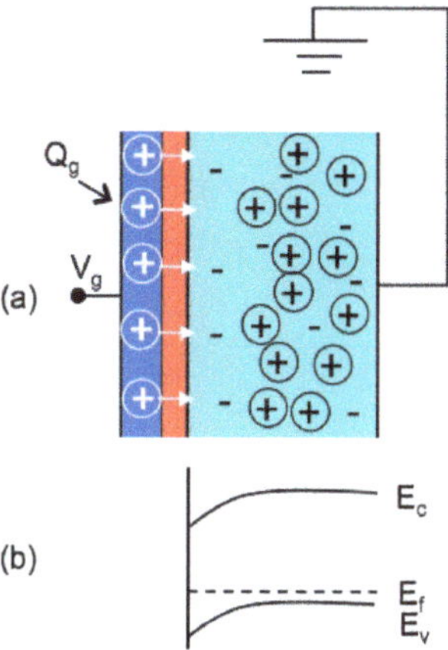

Figure 6.8: Schematic representation of (a) a basic metal-oxide-semiconductor (MOS) structure showing that increasing the voltage extends the depletion region further into the device, and (b) associated band structure of the p-type silicon layer.

As can be seen from Figure 6.8 the electric field now extends into the semiconductor. When there is an electric field, there is a shift in potential, which is represented in the band diagram by bending the bands. Bending the bands down (as we should moving towards positive charge) causes the valence band to pull away from the Fermi level near the surface of the semiconductor. The density of holes in terms of E_v and E_f is determined by,

$$p = N_v e^{-\frac{E_f - E_v}{kT}}$$

and indicates a *depletion region* (region with almost no holes) near the region under the gate. Once E_f - E_v gets large with respect to kT, the negative exponent causes p→0.

The electric field extends further into the semiconductor, as more negative charge is uncovered and the bands bend further down. But now we have to recall the electron density equation, which tells us how many electrons we have:

$$n = N_c e^{-\frac{E_c - E_f}{kT}}$$

Figure 6.9 reveals that with this much band bending the conduction band edge (E_c), and the Fermi level (E_f) are starting to get close to one another (at least compared to kT), which means that n, the electron concentration, should soon start to become significant. The situation in Figure 6.9, represented a condition called *threshold*, and the gate voltage at this point is called the *threshold voltage* (V_T).

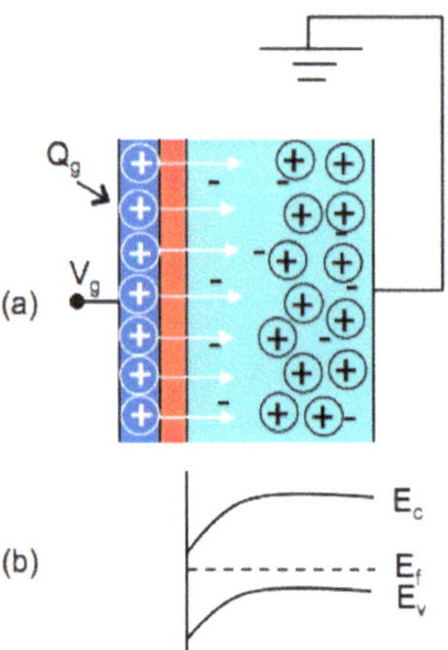

Figure 6.9: Schematic representation of (a) a basic metal-oxide-semiconductor (MOS) structure at threshold, where E_f is close to E_c, and (b) associated band structure of the p-type silicon layer.

Although increase V_g above V_T (Figure 6.10) results in more positive charge at the gate, the depletion region no longer moves back into the substrate. Instead electrons start to appear under the gate region, and the additional electric field lines terminate on these new electrons, instead of on additional acceptors. We have created an *inversion layer* of electrons under the gate, and it is this layer of electrons, which we can use to connect the two n-type regions in our initial device.

Where did these electrons come from? We do not have any donors in this material, so they cannot come from there. The only place from which electrons could be found would be through thermal generation, since in a semiconductor, there are always a few electron hole pairs being generated by thermal excitation at any given time (Figure 1.6). Electrons that get created in the depletion region are caught by the electric field and are swept over to the edge by the gate. In a real MOS device, we have the two n-regions, and it is easy for electrons from one or both to 'fall' into the potential well under the gate, and create the inversion layer of electrons.

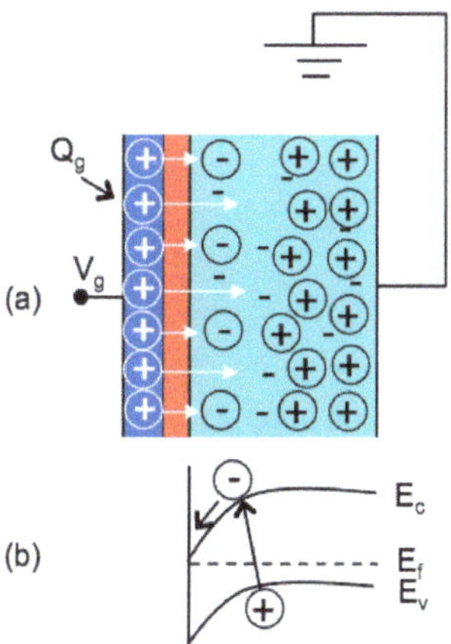

Figure 6.10: Schematic representation of (a) a basic metal-oxide-semiconductor (MOS) structure at *inversion* forming electrons under the gate, and (b) associated band structure of the p-type silicon layer.

Metal-oxide-semiconductor field-effect transistor (MOSFET)

The *metal-oxide-semiconductor field-effect transistor* (MOSFET), also known as the *metal-oxide-silicon* (MOS) transistor, is a type of insulated-gate field-effect transistor that is fabricated by the controlled oxidation of silicon.

Figure 6.11 shows the starting point of MOSFET fabrication starting with p-type doped silicon, into which are imbedded two n-type regions. To each of the n-type regions is attached a contact (wire), and connect a potential between them (Figure 6.11). However, no current (I) will flow through this structure because the n-p junction on the RHS is reverse biased, i.e., the positive lead from the battery going to the n-side of the p-n junction. If we attempt to remedy this by turning the potential around, we will now have the LHS junction reverse biased, and again, no current will flow. In order for current to flow between the 2 terminals it is necessary to form a layer of n-type material between one n-region and the other. This will then connect them together, and we can run current in one terminal and out the other.

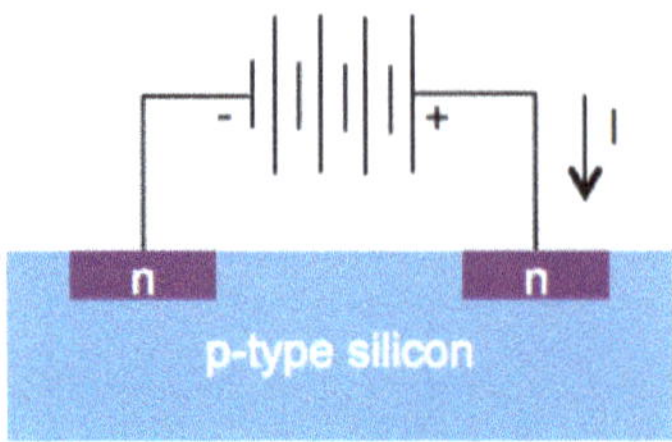

Figure 6.11: The starting point of a field effect transistor (FET).

In order to create a *channel* for conduction the first step is to grow a layer of oxide (SiO_2) on top of the silicon by placing the wafer in an oven under an oxygen atmosphere, and heated to 1100 °C. The result is a high-quality insulating SiO_2 layer on top of the silicon. On top of the oxide layer a conductor is deposited: the *gate*. Traditionally, the gate would have been a layer of metal such as aluminum; hence the *metal-oxide-silicon* name, however, it is common to use a heavily doped layer of polycrystalline silicon deposited from the reduction of a gas, such as silane (SiH_4),

$$SiH_{4(g)} \rightarrow Si_{(s)} + 2\,H_{2(g)}$$

The deposited silicon is polycrystalline because it is deposited on top of the amorphous oxide, which does not provide a suitable substrate for epitaxy and single crystal.

Figure 6.12 now shows both the added oxide layer and *gate* structure, and a second source of a potential so that we can invert the region under the gate (c.f., Figure 6.10) and connect the two n-regions together. The n-region connected to the negative side of the voltage is the *source*, and the other one the *drain*. The source and also the p-type substrate are grounded and potentials are placed between the gate and the source (V_{gs}) and between the drain and the source (V_{ds}).

Figure 6.13 provides a perspective view of the active regions of the MOSFET minus the gate and oxide. Applying a voltage greater than V_T to the gate creates an n-type region, called the *channel*, which connects the *source* and *drain*. We will assume that the channel region is L long and W wide, as shown in Figure 6.13.

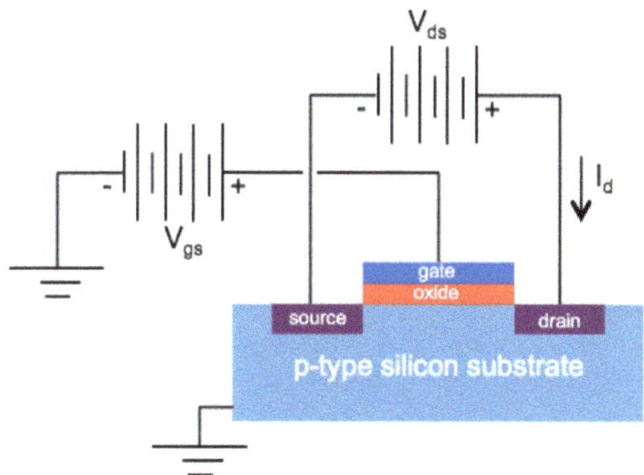

Figure 6.12: A schematic representation of biasing a MOSFET. Note the layer thicknesses are not to scale.

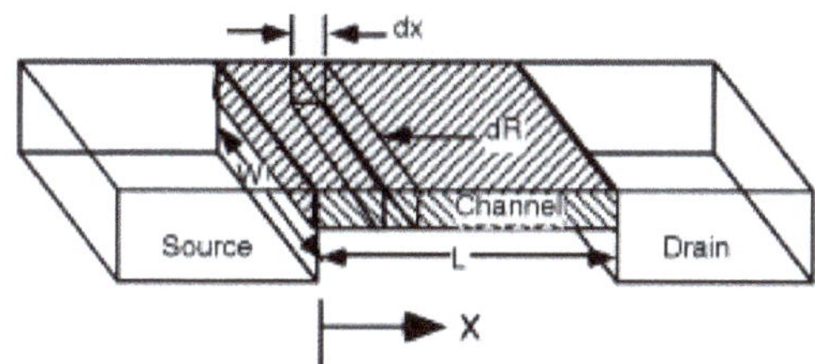

Figure 6.13: A schematic representation of the inversion channel and its resistance for a MOSFET. The substrate, oxide and gate contact are not shown and the layer thicknesses are not to scale.

The resistance $d(R)$ of a small section of the *channel*, that is $d(x)$ in length, is given by,

$$d(R) = \frac{dx}{\sigma_s W}$$

Normally, for the cross-sectional area resistance would be defined by σ in the denominator and an area (A) for the channel; however, it is very hard to determine the cross sectional area of the channel. The electrons that form the inversion layer crowd into a very thin sheet of surface charge, which really has little or no thickness or penetration into the substrate.

On the other hand we consider a surface conductivity, σ_s,

$$\sigma_s = \mu_s Q_{chan}$$

where μ_s is a surface mobility (cm^2/V.sec) that is the quantity which represented the proportionality between the average carrier velocity and the electric field:

$$\bar{v} = \mu E$$

$$\mu = \frac{q\tau}{m}$$

The surface mobility has to be measured for a given system (*ca.* 300 cm^2/V.sec for silicon. Q_{chan} is the *surface charge density* (or *channel charge density*) and has units of C/cm^2. This is like a sheet of charge, which is different from the bulk charge density. Note that:

$$\frac{cm^2}{Volt\ sec}\ \frac{Coulombs}{cm^2}\ =\ \frac{\frac{Coul}{sec}}{Volt}$$

$$=\ \frac{I}{V}$$

$$=\ mhos$$

For any given gate voltage (V_{gs}) the surface charge density in the channel (Q_{chan}) is given simply as:

$$Q_g = c_{ox} V_{gs}$$

However, until the gate voltage (V_{gs}) is greater than V_T there are no mobile electrons created under the gate (c.f., Figure 6.10), and a depletion region is built up (c.f., Figure 6.8). Q_T is defined as the charge on the gate necessary to get to threshold:

$$Q_T = c_{ox} V_T$$

Any charge added to the gate above Q_T, is matched by charge in the channel (Q_{chan}):

$$Q_{\text{channel}} = Q_g - Q_T$$

$$Q_{\text{chan}} = c_{\text{ox}} \left(V_g - V_T \right)$$

The channel resistance is therefore defined as:

$$d(R) = \frac{d(x)}{\mu_s c_{\text{ox}} \left(V_{\text{gs}} - V_T \right) W}$$

As may be seen from Figure 6.12, the current I_d is defined as flowing into the drain. This current flows through the channel, and hence through the incremental resistance, d(R), creating a voltage drop d(V), where V_c is the channel voltage,

$$
\begin{aligned}
d(V_c(x)) &= I_d\, d(R) \\
&= \frac{I_d\, d(x)}{\mu_s c_{\text{ox}} \left(V_{\text{gs}} - V_T \right) W}
\end{aligned}
$$

Integration of this equation will provide the value for the entire channel. The voltage on the channel, $V_c(x)$, goes from 0 on the left to V_{ds} on the right (Figure 6.13). At the same time, x is going from 0 to L. Thus the limits of integration will be 0 and V_{ds} for the voltage integral $d(V_c(x))$ and from 0 to L for the x integral d(x):

$$\int_0^{V_{\text{ds}}} \mu_s c_{\text{ox}} \left(V_{\text{gs}} - V_T \right) W\, d\,V_c = \int_0^{L} I_d\, d\,x$$

This is rearranged to provide I_d as a function of applied voltages:

$$I_d L = \mu_s c_{\text{ox}} W \left(V_{\text{gs}} - V_T \right) V_{\text{ds}}$$

Division by L, gives an expression for the drain current (I_d) in terms of the drain-source voltage (V_{ds}), the gate voltage (V_{gs}), and the physical attributes of the MOS transistor:

$$I_d = \left(\frac{\mu_s c_{\text{ox}} W}{L} \left(V_{\text{gs}} - V_T \right) \right) V_{\text{ds}}$$

Light emitting diode (LED)

When the electron falls down from the conduction band and fills in a hole in the valence band, there is a loss of energy. The question is; where does that energy go?

In silicon, the answer is not very interesting. Silicon is what is known as an *indirect band-gap* material. What this means is that as an electron goes from the bottom of the conduction band to the top of the valence band, it must also undergo a significant change in momentum. Whenever something changes state, we must still conserve not only energy, but also momentum. In the case of an electron going from the conduction band to the valence band in silicon, both of these things can only be conserved if the transition also creates a quantized set of lattice vibrations, called *phonons*, or heat. Phonons posses both energy and momentum, and their creation upon the recombination of an electron and hole allow for complete conservation of both energy and momentum. All of the energy that the electron gives up in going from the conduction band to the valence band (1.1 eV) ends up in phonons, which is another way of saying that the electron heats up the crystal.

In a class of materials called *direct band-gap* semiconductors, the transition from conduction band to valence band involves essentially no change in momentum. Photons, it turns out, possess a fair amount of energy (several eV/photon in some cases) but they have very little momentum associated with them. Thus, for a direct band gap material, the excess energy of the electron-hole recombination can either be taken away as heat, or more likely, as a photon of light. This radiative transition then conserves energy and momentum by giving off light whenever an electron and hole recombine. This gives rise to the light emitting diode (LED). Emission of a photon in an LED is shown schematically in Figure 6.14.

It was Planck who postulated that the energy of a photon was related to its frequency by a constant, which was later named after him. For the frequency of oscillation (v) the energy of the photon is just given by,

$$E = hv$$

where h is Planck's constant, which has a value of 4.14×10^{-15} eV.sec.

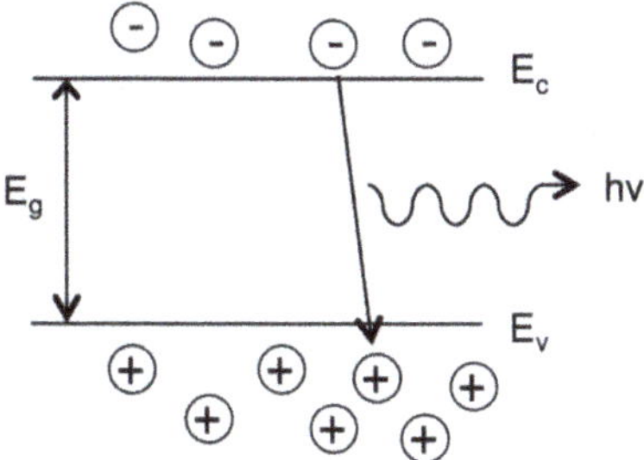

Figure 6.14. Schematic representation of the radiative recombination in a direct band-gap (E$_g$) semiconductor such as GaAs.

When we talk about light it is conventional to specify its wavelength (λ) instead of its frequency. Visible light has a wavelength on the order of nanometers, e.g., red $\approx$ 600 nm, green $\approx$ 500 nm and blue $\approx$ 450 nm region. The relationship between wavelength and band gap (E$_g$) can be derived from $c = \lambda v$, where c is the speed of light (3 x 10^3 m/sec or 3 x 10^{17} nm/sec):

$$\lambda\,(\text{nm}) \;=\; \frac{hc}{E(\text{eV})}$$
$$\;=\; \frac{1242}{E(\text{eV})}$$

Thus, a semiconductor with a 2 eV band-gap should give off light at about 620 nm (i.e., in the red). A 3 eV band-gap material would emit at 414 nm, in the violet. The various semiconductor materials, used for important light emitting diodes (LEDs), are shown in Figure 6.15 with their emitted color, along with the response of the human eye (Figure 6.15).

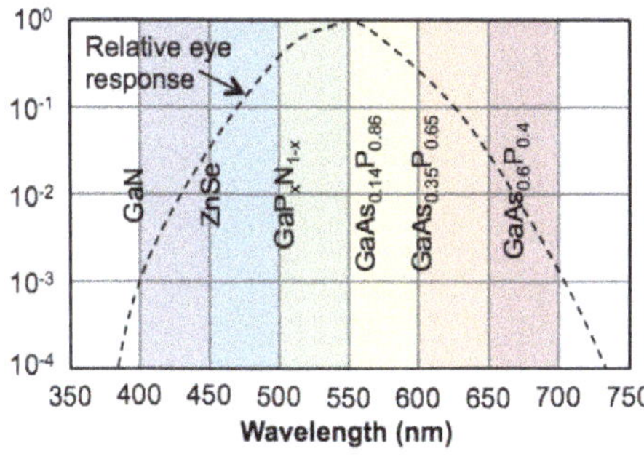

Figure 6.15: Relative response of the human eye to various colors and the direct band gap semiconductors.

It is worth noting that a number of the important LEDs are based on the GaAsP system. GaAs is a direct band-gap semiconductor with a band gap of 1.42 eV (in the infrared). GaP is an indirect band-gap material with a band gap of 2.26 eV (550 nm = green). Both As and P are group V (15) elements and their gallium compounds are isomorphous (Table 2.3). We can replace some of the As with P in GaAs and make a mixed compound semiconductor $GaAs_{1-x}P_x$. When the mole fraction of phosphorous is less than about 0.45 the band gap is direct, and so it is possible to 'engineer' the desired color of LED required by simply growing a crystal with the proper phosphorus concentration. The properties of the GaAsP system are shown in Figure 6.16.

For the GaAsP system, there are actually two different band gaps, as shown in Figure 6.16. One is a *direct* gap (no change in momentum) and the other is *indirect*. In GaAs, the direct gap has lower energy than the indirect one (Figure 6.17a) and so the transition is a radiative one. As we start adding phosphorous to the system, both the direct and indirect band gaps increase in energy; however, the direct gap energy increases faster with phosphorous fraction than does the indirect one. At a mole fraction (x) of about 0.45, the gap energies cross over and the material goes from being a direct gap semiconductor to an indirect gap semiconductor (Figure 6.17c). At $x = 0.35$ the band gap is about 1.97 eV (630 nm), and so we would only expect to get light up to the red using the GaAsP system for making LED's. Fortunately, the addition of nitrogen dopant to the GaAsP system introduces a new level in the system (Figure 6.17c). An electron can go from the indirect conduction band (for a mixture with a mole fraction greater than 0.45) to the nitrogen site, changing its momentum, but not its energy. It can make a direct transition to the valence band, and light with colors all the way to the green became possible.

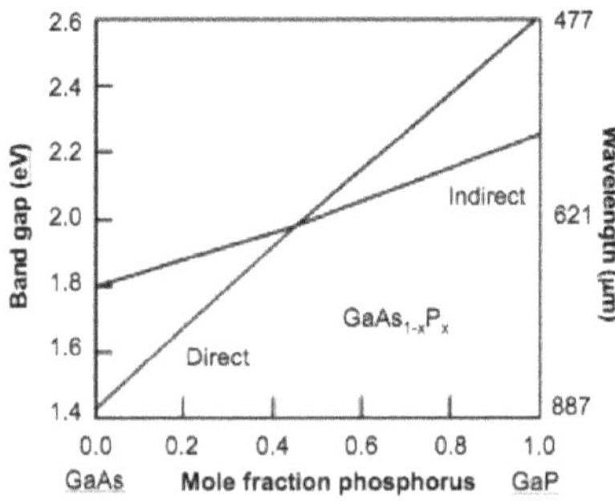

Figure 6.16: Band gap for the GaAsP system as a function of composition.

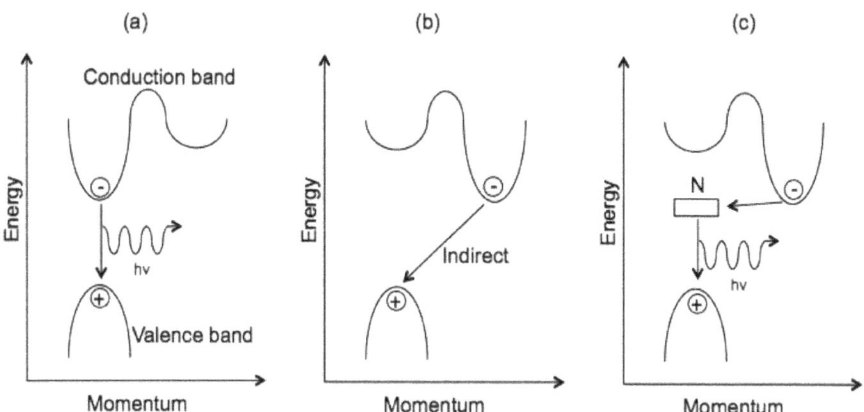

Figure 6.17: Schematic representation of the relationship between direct and indirect band gaps for GaAs$_{1-x}$P$_x$ with (a) x < 0.45, (b) x > 0.45, and (c) the effect of addition of a nitrogen recombination center x > 0.45.

For LEDs with wavelengths shorter than the green the GaAsP system must be replaced with more suitable materials. A compound semiconductor made from the II-VI (12-16) elements Zn and Se make up one promising system, and several research groups have successfully made blue and blue-green LEDs from ZnSe. SiC is another (weak) blue emitter, however, it was the successfully fabricated a blue LED using the III-V material GaN that significantly changed LED technology. Adding blue to the already working green and red LED's completes the set of 3 primary colors necessary for a full-color flat panel display. Furthermore, using a blue LED or laser in a CD ROM more than quadrupled its data capacity, as bit diameter scales as λ, and hence the area as λ^2.

Polymer light emitting diodes

Electroluminescent (EL) from conjugated polymers was first reported in 1990 when a layer of poly(*para*-phenylenevinylene) (PPV, Figure 6.18) was sandwiched between layers of indium tin oxide (ITO) and aluminum. When this device is under a 14 V DC bias, the PPV emits a yellowish-green light with a quantum efficiency of 0.05%. This research attracted a lot of attention, because the potential that polymer light emitting diodes (LEDs) could be inexpensively mass-produced into large area display area. The processing steps in making polymer LEDs are readily scalable. The industrial coating techniques is well developed to mass-produce polymer layers of 100 nm thickness, and the device could be patterned onto large surface area by pixelation of metal.

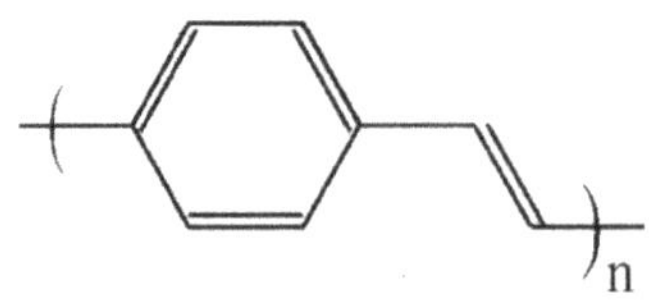

Figure 6.18: Structure of poly(*para*-phenylenevinylene) (PPV).

Since the initial discovery, and increasing amount of researches has been performed, and significant progress has been made. In 1990 the polymer LED only emitted yellowish green color, now the emission color ranged from deep blue to near infra red. The efficiency of the multi-layer polymer LED even reached a quantum efficiency of >4% and the operating voltage has been reduced significantly. In term of efficiency, color selection and operating voltage, polymer LEDs have attained adequate levels for commercialization. But there are reliability problems that are symptomatic of any organic devices.

A schematic diagram of a polymer LED is shown in Figure 6.19. A polymer LED can be divided into three different components:
- an *anode* that is the hole supplier, made of metal of high working function. Examples of the common anode are indium tin oxide (ITO), gold etc. The anode is usually transparent so that light can be emitted through.
- a *cathode* that is the electron supplier, made of metal of low working function. Examples of the common cathode are aluminum or calcium.
- a *polymer* made of conjugated polymer film with thickness of 100 nm.

When a polymer LED is under a direct current (dc) bias, holes are injected from the anode (ITO) and electrons are injected from the cathode (aluminum). Under the influences of the electrical field, the electrons and holes will migrate toward each other. When they recombine in the conjugated polymer layer, bound excited states (*excitons*) will be formed. Some of the excitons (*singlets*) then decays in the conjugated polymer layer to emit light through the transparent substrates (glass). The emission color will be depended on the energy gap of the polymers. The π electron is not completely delocalized over the entire polymer chain. Instead there are alternate region in the polymer chain that has a higher electron density (Figure 6.20a). The chain length of this region is about 15-20 multiple bonds. The emission color can be controlled by tuning the band gap (Figure 6.20b). It shows that bond alternation limits the extent of delocalization. Table 6.1 summarizes the structure and emission color of some common conjugated polymers.

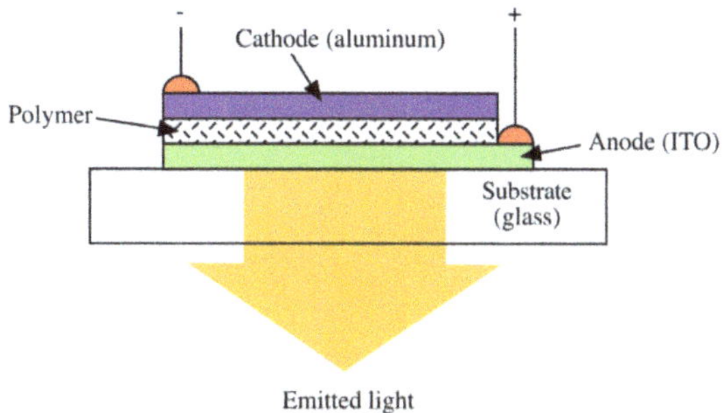

Figure 6.19: Schematic set-up of polymer LED.

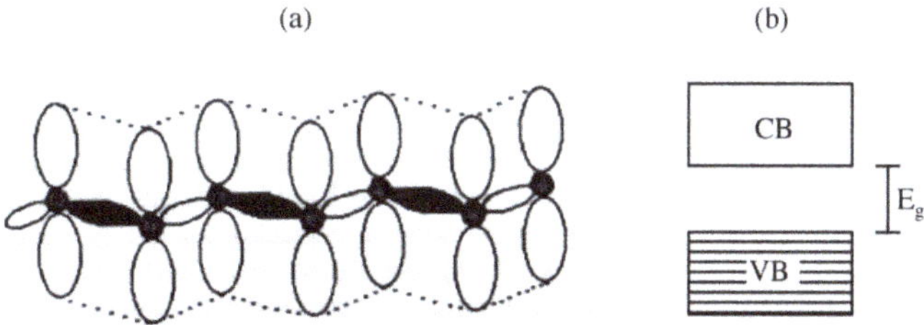

Figure 6.20: Alternation of bond lengths along a conjugated polymer chain (a) results in a material with properties of a large band gap semiconductor (b) where CB is the conductive band gap, and VB is the valence band, and E_g is the band gap.

Efficiency for any LED is defined:

$$n_{ext} = n_{esc} \times n_{int}$$

where n_{ext} is the external quantum efficiency, n_{int} is the internal efficiency (represents the fraction of injected carrier, usually electron, that is converted to photon), and n_{esc} is the escape efficiency (represent fraction of photons that can reach to the outside).

Poly-mer	Chemical name	Structure	π-π* energy gap (eV)	Emission peak (nm)
PA	*trans*-polyacetylene		1.5	600
PDA	polydiacetylene		1.7	
PPP	poly(*para*-phenylene)		3.0 (red)	465
PPV	poly(*para*-phenylene-vinylene)		2.5 (green)	565
RO-PPV	poly(2,5-dialkoxy-p-phenly-enevinylen)		2.2 (blue)	~580
PT	polythiophene		2.0 (red)	
P3AT	poly(3-alkythiophene)		2.0 (red)	690
PTV	poly(2,5-thio-phenevinylene)		1.8	
PPy	polypyrrole		3.1	
PAni	polyaniline			3.2

Table 6.1: Example of common conjugated polymers.

The most common way to improve the internal efficiency is to balance the number of electrons and holes, which arrives at the polymer layer. Originally, there are more holes than electron that arrive of the polymer layer because conjugated polymers have a higher electron affinity, and as a consequence will favor the transport of hole than electron. There are two ways to maintains the balance:

- match the work function of electrode with the electron affinity and ionization potential of the polymer.
- tune the polymer's electron affinity and ionization potential to match the work function of the electrode.

The escape efficiency is also important because a polymer LED is made up of layers of materials that have different refractive index, and some of the photon generated from the excition may be reflected at the boundary and trapped inside the device.

Improvement in internal quantum efficiency using low working function cathode

Conjugated polymer is electron rich, the mobility for hole is higher than electron, and more holes will arrive in the polymer layer than electrons. One way to increase the population of the electron is to use a lower working function metal as cathode. Replacement of the aluminum cathode with calcium results in improved internal efficiency by a factor of ten, to 0.1%. This approach is direct and fast but low working function electrode like calcium will be oxidized easily and shorten the devices' life.

Improvement in internal quantum efficiency using multiple polymer layers

A layer of poly[2,5-di(hexyloxy)cyanoterephthalylidene] (CN-PPV, Figure 6.21) is coated on top of PPV to improve the transport and recombination of electron and holes (Figure 6.22).

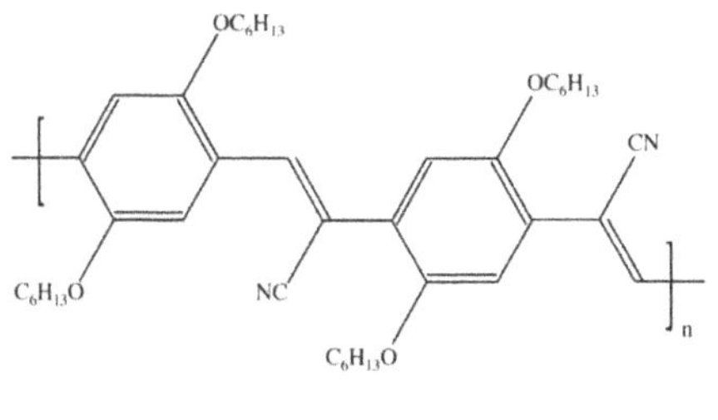

Figure 6.21: Structure of poly[2,5-di(hexyloxy)cyanoterephthalylidene] (CN-PPV).

The nitrile group in the CN-PPV has two effects on the polymer.

- It increases the electron affinity so electrons can travel more efficient from the aluminum to the polymer layer. And metal of relative high working function like aluminum and gold can be now be used as cathode instead of calcium.
- It increases the binding energy of the occupied π and unoccupied π^* state but maintain a similar π-π^* gap. So when the PPV and CN-PPV is placed together, holes and electron will be confined at the heterojunction.

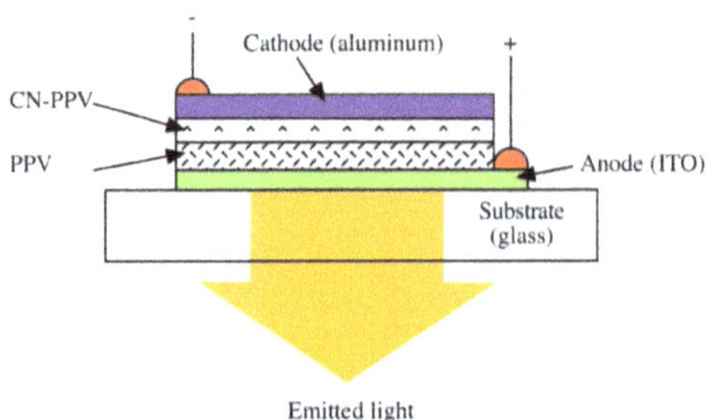

Figure 6.22: Schematic representation of a CN-PPV and PPV multi-layer LED.

Based on this approach, a couple of polymers have been developed or modified to produce the desirable emission color and processing property. The drawback of this method is that desirable properties may not be commentary to each other. For example, in MEH-PPV an alkoxy side group (RO) is introduced to PPV so that it can be dissolved in organic solvent. But the undesirable effect is that MEH-PPV is less thermally stable. Moreover in multiple layers LEDs, different polymer layers have different refractive indices and a fraction of the photons will undergo total internal reflection at the refractive boundaries and cannot escape as light. This problem can be overcome by Febry-Pert microcavity structure.

Improvement in external quantum efficiency using microcavity

Fabry-Perot resonant structures are also used in inorganic LED, and are is based on Fermi's golden rule:

$$K_r \sim |<M>| \, r_{(v)}$$

where M (the matrix element of the perturbation between final and initial states) depends on the nature of the material, and $r_{(v)}$ can be altered by changing the density of various density states, e.g., using a luminescent thin films to select certain value of V.

In building a microcavity for a polymer LED, the polymer is placed between two mirrors. (Figure 6.23), in which one of the mirrors is made up of aluminum, the other mirror (a Bragg Mirror) is form by epitaxial multilayer stacks of Si_xN_y and SiO_2.

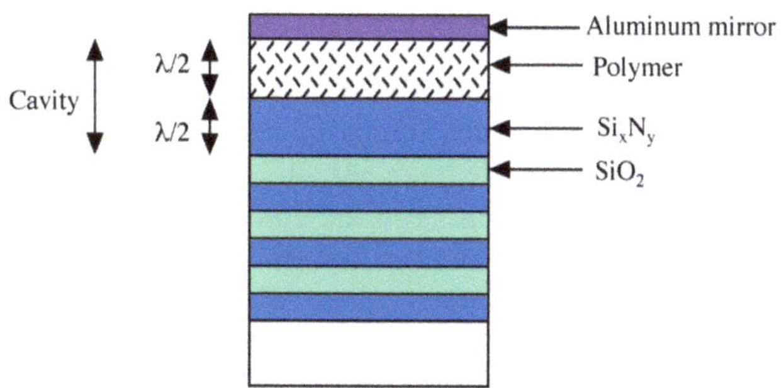

Figure 6.23: Schematic set-up of micro-cavity.

Improvement in internal quantum efficiency: doping of polymer

Doping is a process that creates carrier by purposely introducing impurities and is very popular method in the semiconductor industry. However, this technique was not used in polymer LED until 1995, when a co-polymer polystyrene-poly(3-hexylthiphene) (PS-P3HT) was doped with $FeCl_3$. Doping of MEH-PPV with iodine has improved the efficiency by 200% and the polymer LED can be operated under both forward and reverse bias (Table 6.2). The doping is accomplished by mixing 1 wt.% MEH-PPV with 0.2 wt.% I_2. The molar ratio of MEH-PPV to I_2 is 5:1. That is a huge "doping " ratio when you compare the doping concentration in the semiconductor.

	Un-doped	**Doped**
Turn on voltage (V)	10	Foreword 5, reversed 12
External efficiency (%)	4×10^{-4}	8×10^{-3}

Table 6.2: Results of iodine doping of an Al/MEH-PPV/ITO-based LED.

Polymer LEDs on a silicon substrate: an application advantage over inorganic LEDs

In the initial research polymer LEDs were in direct competition with the inorganic LEDs and tried to achieve the existing LED standard. This is a difficult task since polymer LEDs have a lower long term stability; however, there are some applications in which polymer LEDs have a clear advantage over their more traditional inorganic analogs. One of these is to incorporate LEDs with the silicon integrated circuits for inter-chip communication.

It is difficult to build inorganic LEDs on a silicon substrate, because of the thermal stress developing between the inorganic LED (usually a III-V based device) and the silicon interface. But polymer LEDs offer a solution, since polymers can be easily spin-coated on the silicon. The operating voltage of polymer LED is less than 4 V, and the turn on voltage can be as low as 2 V. Together with a switching time of less than 50 ns, make polymer LED a perfect candidate.

Reliability and degradation of polymer LEDs

In terms of the efficiency, color selection, and driving voltage, polymer LEDs have attained adequate level for commercialization. However, the device lifetime is still far from satisfactory. Research into understanding the reliability and degradation mechanisms of polymer LEDs has generally been divided into two areas:
- photo-degradation of polymer.
- interface degradation.

Solid state laser

A laser is a device that emits light through a process of optical amplification based on the *stimulated emission* of *electromagnetic radiation*. The term *laser* originated as an acronym for "light amplification by stimulated emission of radiation".

What is the difference between an LED and a solid-state laser? After all, both devices operate on the same principle of having excess electrons in the conduction band of a semiconductor, and arranging it so that the electrons recombine with holes in a radiative fashion, giving off light in the process. What is different about a laser? In an LED, the electrons recombine in a random and unorganized manner. They give off light by what is known as *spontaneous emission*, which simply means that the exact time and place where a photon comes out of the device is up to each individual electron, and things happen in a random way. There is another way in which an excited electron can emit a photon. If a field of light (or a set of photons) happens to be passing by an electron in a high-energy state, that light field can induce the electron to emit an additional photon through a process called stimulated emission. The photon field *stimulates* the electron to emit its energy as an additional photon, which comes out *in phase with the stimulating field*. This is the big difference between *incoherent light* (what comes from an LED or a flashlight) and *coherent light*, which comes from a laser. With coherent light, all of the electric fields associated with each phonon are all exactly in phase. This coherence is what enables us to keep a laser beam in tight focus, and to allow it to travel a large distance without much divergence or spreading out.

So how do we restructure an LED so that the light is generated by stimulated emission rather than spontaneous emission? Firstly, the fundamental structure of a solid-state laser is a *heterostructure*: a sandwich of different materials with different characteristics. In this case, two wide band-gap regions are placed around a region with a narrower band gap. The most important system is the AlGaAs/GaAs system. A band diagram for such a set up is shown in Figure 6.24. AlGaAs (pronounced "Al-Gas") has a larger band-gap than GaAs. The potential *well* formed by the GaAs means that the electrons and holes will be confined there, and all of the recombination will occur in a very narrow strip. This greatly increases the chances that the carriers can interact, but we still need some way for the photons to behave in the proper manner.

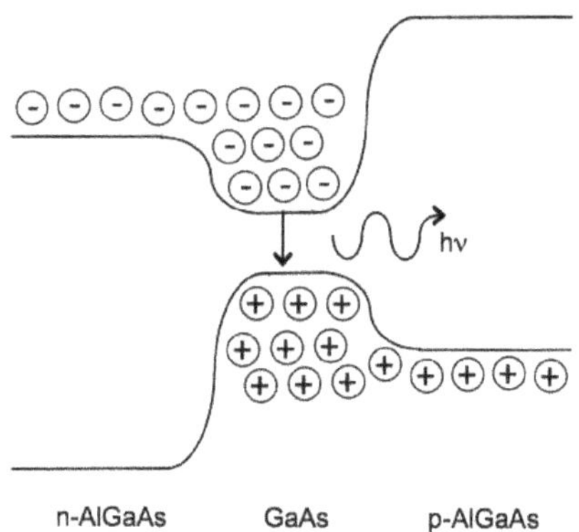

Figure 6.24: A schematic representation of the band diagram for a double heterostructure GaAs/AlGaAs laser.

Figure 6.25 is a diagram of the structure of a typical laser diode in which the active GaAs layer is sandwiched in-between the two heterostructures confinement layers, with a contact on top and bottom. On either end of the device, the crystal has been 'cleaved' or broken along a crystal lattice plane. This results in a shiny 'mirror-like' surface, which will reflect photons. The back surface (which we can not see in Figure 6.25) is also cleaved to make a mirror surface. The other surfaces are purposely roughened so that they do not reflect light.

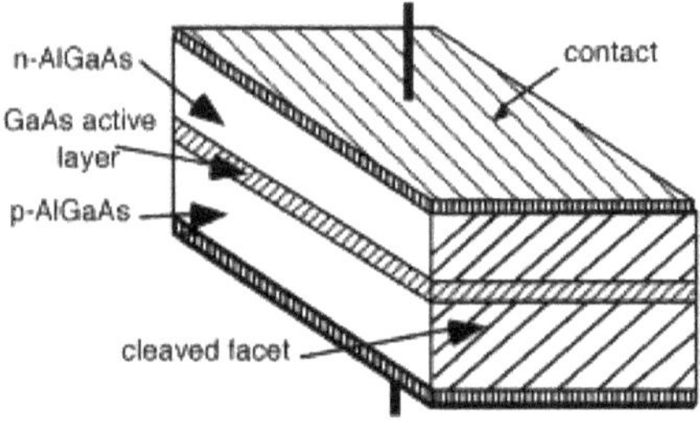

Figure 6.25: A schematic diagram of a typical laser diode.

Figure 6.26 shows the band diagram for the GaAs region of the laser diode in Figure 6.25. The process is initiated by an electron and hole recombining spontaneously, which emits a photon that heads towards one of the mirrored faces (Figure 6.26a). As the photon goes by other electrons; however, it may cause one of them to decay by *stimulated emission* (Figure 6.26b). The two (in phase) photons hit the mirror and are reflected and start back the other way. As they pass additional electrons, they stimulate them into a transition as well (Figure 6.26c), and the optical field within the laser starts to build up. After a bit, the photons get down to the other end of the cavity (right hand side of Figure 6.26). The cleaved facet, while it acts like a mirror, is not a perfect one. Some light is not reflected, but rather 'leaks' though, and so becomes the output beam from the laser (Figure 6.26d). The rest of the photons are reflected back into the cavity and continue to stimulate emission from the electrons, which continue to enter the gain region because of the forward bias on the diode.

In reality, the photons do not move back and forth in a big 'clump' as described here, rather they are distributed uniformly along the gain region. The field within the cavity will build up to the point where the loss of energy by light leaking out of the mirrors just equals the rate at which energy is replaced by the recombining electrons.

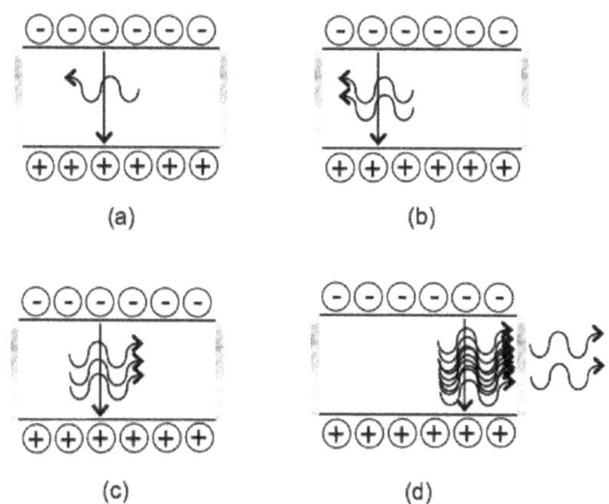

Figure 6.26: Schematic representation of (a – c) the stepwise buildup of a photon field and (d) the output coupling in a diode laser.

Solar cells

A solar cell, or photovoltaic cell, is an electrical device that converts the energy of light directly into electricity by the photovoltaic effect, which is a physical and chemical phenomenon.

Consider the following situation with a normal p-n junction, instead of applying an external voltage, as is done in a LED, the junction is illuminated with light whose photon energy is greater than the band-gap (Figure 6.27a). Instead of recombination, photo-generation of electron hole pairs will occur because the photons excites an electron from the full states in the valence band, and 'kicks' them up into the conduction band, leaving a hole behind. This is similar to the thermal excitation process shown in Figure 1.6. As can be seen from Figure 6.27b, such a process creates excess electrons in the conduction band in the p-side of the diode, and excess holes in the valence band of the n-side. These carriers can diffuse over to the junction, where they will be swept across by the built-in electric field in the depletion region. If the two sides of the diode were connected together with a wire, a current would flow through that wire as a result of the electrons and holes, which move across the junction.

Which way would the current flow? A quick look at Figure 6.27c shows that holes (positive charge carriers) generated on the n-side will 'float' up to the p-side as they go across the junction. Hence, positive current must be coming out of the anode, or p-side of the junction. Likewise, electrons generated on the p-side will fall down the junction potential, and come out the n-side, but since they have negative charge, this flow represents current going into the cathode. The resulting device is a *photovoltaic diode* or *solar cell*.

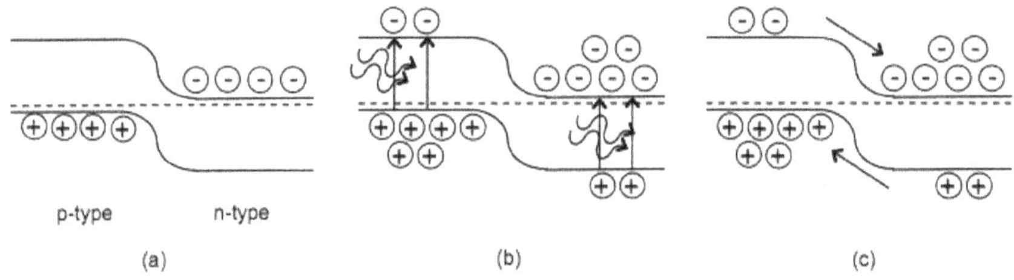

Figure 6.27: A schematic representation of a p-n diode under illumination.

Figure 6.28a is a diagram of what a solar cell would look like schematically. Although this device would appear to be used as a source of energy, but the way it is set up the voltage across the diode is zero, and since power equals I×V, there is no power generated. In order to generate power a load resistor is required, i.e., Figure 6.28b. Now the photo current flowing through the load resistor will develop a voltage, which biases the diode in the forward direction, that will in turn cause current to flow back into the anode.

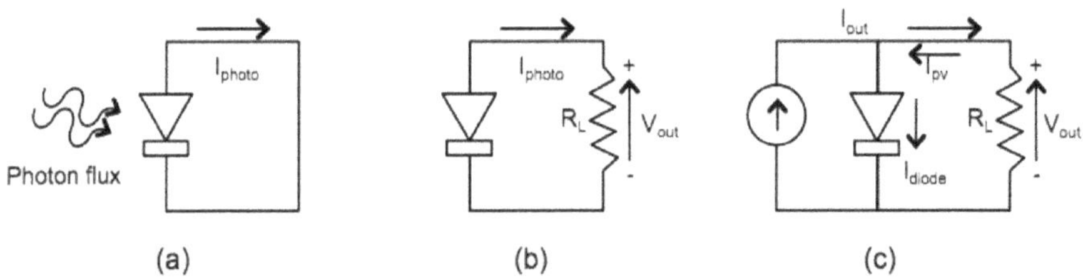

Figure 6.28: Schematic representation of (a) a photovoltaic cell, (b) a photovol-taic cell with a load resistor, and (c) a model of a photovoltaic cell.

The current that arises due to the photon flux can be conveniently represented as a current source (Figure 6.28c). Even though I_{out} is shown coming out of the device, the usual polarity convention that when we define V_{out} as being positive at the top, then we should show the current for the photovoltaic, I_{pv} as current going into the top, see Figure 6.28c. Note that $I_{pv} = I_{diode} - I_{photo}$, so all we need to do is to subtract the two currents; we do this graphically in Figure 6.29. The four quadrants in the I-V plot are numbered. In quadrant I and III, the product of I and V is a positive number, meaning that power is being dissipated in the cell. For quadrant II and IV, the product of $I \times V$ is negative, and so we are getting power from the device. Clearly operation in quadrant IV is desirable. In fact, without the addition of an external battery or current source, the circuit, will only run in the IVth quadrant. Consider adjusting R_L, the load resistor from 0 (a short) to ∞ (an open). With R_L, we would be at point A on Figure 6.29. As R_L starts to increase from zero, the voltage across both the diode and the resistor will start to increase also, and we will move to point B, say. As R_L gets bigger and bigger, we keep moving along the curve until, at point C, where R_L is an open and we have the maximum voltage across the device, but, of course, no current coming out.

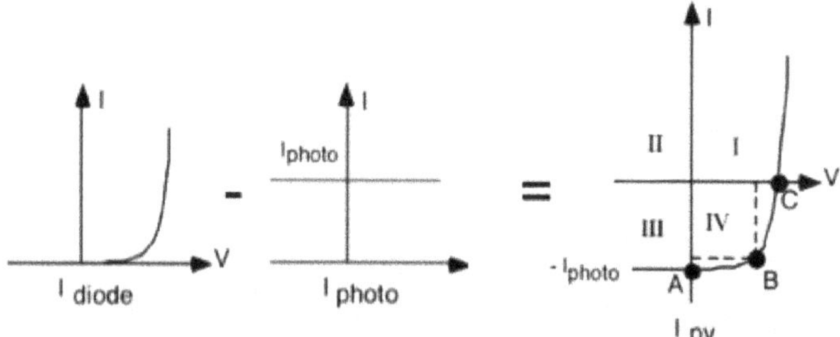

Figure 6.29: Combining the diode and the current source.

Power is VI so at B (Figure 6.29) for instance, the power coming out would be represented by the area enclosed by the two dotted lines and the coordinate axes. Someplace about where I have point B would be where we would be getting the most power out of out solar cell.

Figure 6.30 shows you what a real solar cell would look like. They are usually made from a complete wafer of silicon, to maximize the usable area. A shallow (0.25 µm) junction is made on the top, and top contacts are applied as stripes of metal conductor as shown. An anti-reflection (AR) coating is applied on top of that, which accounts for the bluish color, which a typical solar cell has (Figure 6.31).

The solar power flux on the earth's surface is about 1 kW/m^2 or 100 mW/cm^2. So if we made a solar cell from a 4" diameter wafer it would have an area of about 81 cm^2 and so would be receiving a flux of about 8.1 Watts. Typical terrestrial cell efficiencies run from about 10%-15%. This means that we will get about 1.2 Watts out from a single cell. Looking at B on Figure 6.30 we could guess that V_{out} will be about 0.5 to 0.6 V, thus we could expect to get maybe around 2.5 amps from a 4" wafer at 0.5 volts with 15% efficiency under the illumination of one sun.

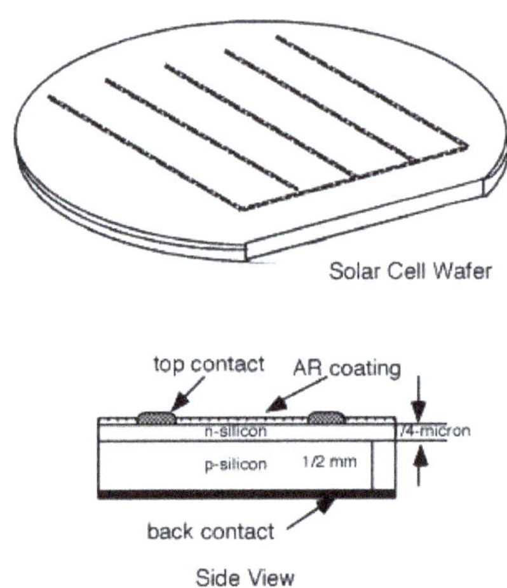

Figure 6.30: A schematic diagram of solar cell.

Figure 6.31: A solar cell showing the blue tint due to the anti-reflective coating.

Chapter 7: Doping

In semiconductor production, *doping* is the intentional introduction of impurities into an intrinsic semiconductor for the purpose of modulating its electrical and optical properties. The doped material is referred to as an extrinsic semiconductor. A semiconductor doped to such high levels that it acts more like a conductor than a semiconductor is referred to as a degenerate semiconductor.

Starting with a polished wafer then how do we get an integrated circuit? We will focus on the *complementary metal oxide semiconductor* (CMOS) process. Starting with a wafer that has some level of doping as a consequence of the crystal growth so that it has a background concentration of acceptors, i.e., it is p-type. In a CMOS logic device the first thing we need to build is a region of n-type doping. In order to do this, we need some way in which to introduce additional (and specific) impurities into the semiconductor. There are several ways to do this, but current technology relies mainly on a technique called *ion implantation*. An ion implanter uses a dopant source gas, ionizes it, and drives the ions into the wafer (Figure 7.1).

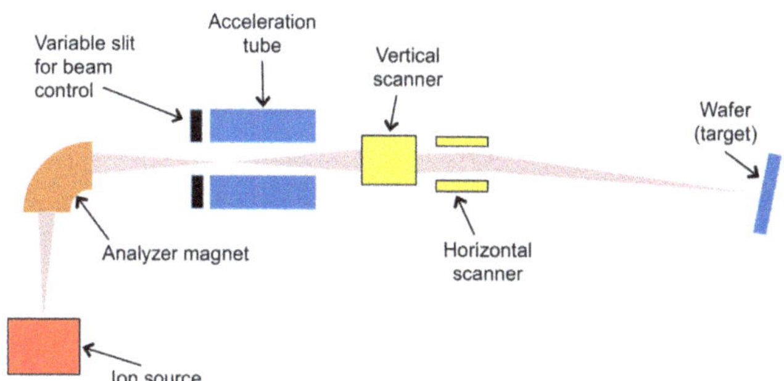

Figure 7.1: Schematic diagram of an ion implantation system.

The dopant gas is ionized and the resultant charged ions are accelerated through a magnetic field, where they are mass analyzed. The vertical magnetic field causes the beam of ions to spread out, according to their mass. A thin aperture selects the ions of interest, and lets them pass, blocking all the others. This makes sure that only the desired ions are implanted, and in fact, it is even possible to select a specific proper isotope. The ionized atoms are then accelerated through several tens to hundreds of kV, and then deflected by an electric field, much like in an oscilloscope CRT. In fact, most of the time the ion beam is rastered across the surface of the silicon wafer. The ions strike the silicon wafer and pass into its interior. A measurement of the current flow in the system and its integral is a measure of how much dopant was deposited into the wafer. This is usually given in terms of the number of dopant atoms/cm^2 to which the wafer has been exposed.

After the atoms enter the silicon, they interact with the lattice, creating defects, and slowing down until finally they stop. Typical atomic distributions, as a function of implant voltage are show in Figure 7.2 for implantation into amorphous silicon. When implantation is done on single crystal material, channeling, the improved mobility of an ion can effect the impurity distribution significantly. Just slight changes of less than a degree can make big differences in how the impurity atoms are finally distributed in the wafer. Usually, the operator of the implant machine purposely tilts the wafer a few degrees off normal to the beam in order to arrive at more reproducible results.

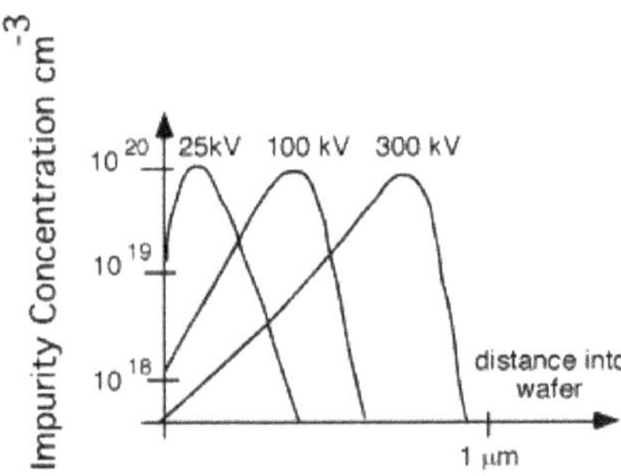

Figure 7.2: Implant distribution with acceleration energy

Impinging 100 kV ions at crystalline silicon causes damage to the crystal lattice. Not only that, but just having, say boron, in your wafer does not mean you are going to have holes (p-type dopant). For the boron to become electrically active, that is to act as an acceptor, it has to reside on a silicon lattice site. Even if the boron atom does, somehow, end up on an actual lattice site the many defects, which have been created, will act as deep traps. Thus, the hole that is formed will probably be caught at a trap site and will not be able to contribute to electrical conductivity in the wafer anyway. Annealing is used to fix this situation, with typical temperatures of 500 - 1000 °C for 10 - 30 minutes.

Chapter 8: Oxidation of Silicon

In the fabrication of integrated circuits (ICs), the oxidation of silicon is essential, and the production of superior ICs requires an understanding of the oxidation process and the ability to form oxides of high quality. Silicon dioxide has several uses:

- serves as a mask against implant or diffusion of dopant into silicon,
- provides surface passivation,
- isolates one device from another (dielectric isolation),
- acts as a component in MOS structures,
- provides electrical isolation of multi-level metallization systems.

Methods for forming oxide layers on silicon have been developed, including thermal oxidation, wet anodization, chemical vapor deposition (CVD), and plasma anodization or oxidation. Generally, CVD is used when putting the oxide layer on top of a metal surface, and thermal oxidation is used when a low-charge density level is required for the interface between the oxide and the silicon surface.

Silicon's surface has a high affinity for oxygen and thus an oxide layer rapidly forms upon exposure to the atmosphere. The chemical reactions, which describe this process are:

$$Si_{(s)} + O_{2(g)} \rightarrow SiO_{2(s)}$$

$$Si_{(s)} + 2\,H_2O_{(g)} \rightarrow SiO_{2(s)} + 2\,H_{2(g)}$$

In the first reaction a dry process is utilized involving oxygen gas as the oxygen source and the second reaction describes a wet process, which uses steam. The dry process provides a 'good' silicon dioxide but is slow and mostly used at the beginning of processing. The wet procedure is problematic in that the purity of the water used cannot be guaranteed to a suitable degree. This problem can be easily solved using a pyrogenic technique, which combines hydrogen and oxygen gases to form water vapor of very high purity. Maintaining reagents of high quality is essential to the manufacturing of integrated circuits and is a concern which plagues each step of this process.

The formation of the oxide layer involves shared valence electrons between silicon and oxygen, which allows the silicon surface to rid itself of *dangling bonds*, such as lone pairs and vacant orbitals, Figure 8.1. These vacancies create mid-gap states between the valence and conduction bands, which prevent the desired band gap of the semiconductor. The Si-O bond strength is covalent (strong), and so can be used to achieve the loss of mid-gap states and *passivate* the surface of the silicon.

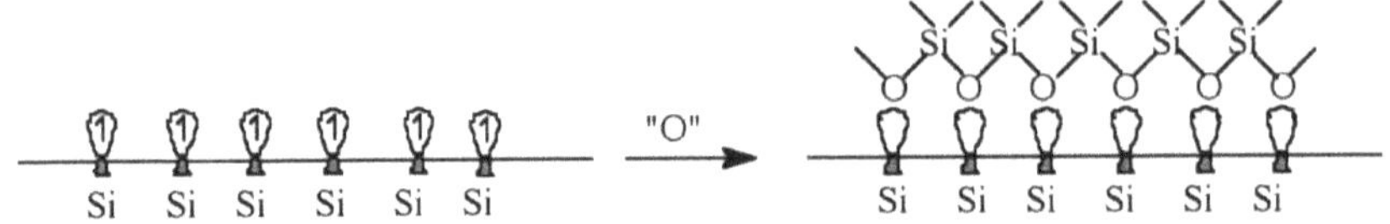

Figure 8.1: Removal of dangling bonds by oxidation of surface.

The oxidation of silicon occurs at the silicon-oxide interface and consists of four steps:

1. Diffusive transport of oxygen across the diffusion layer in the vapor phase adjacent to the silicon oxide-vapor interface.
2. Incorporation of oxygen at the outer surface into the silicon oxide film.
3. Diffusive transport across the silicon oxide film to its interface with the silicon lattice.
4. Reaction of oxygen with silicon at this inner interface.

As the Si-SiO$_2$ interface moves into the silicon its volume expands and based upon the densities and molecular weights of Si and SiO$_2$, 0.44 Å Si is used to obtain 1.0 Å SiO$_2$.

Pre-oxidation cleaning

The first step in oxidizing a surface of silicon is the removal of the native oxide, which forms due to exposure to open air. This may seem redundant to remove an oxide only to put on another, but this is necessary since uncertainty exists as to the purity of the oxide, which is present. The contamination of the native oxide by both organic and inorganic materials (arising from previous processing steps and handling) must be removed to prevent the degradation of the essential electrical characteristics of the device. A common procedure uses an H$_2$O-H$_2$O$_2$-NH$_4$OH mixture which removes the organics present, as well as some group I and II (1 and 2) metals. Removal of heavy metals can be achieved using an H_2O-H_2O_2-HCl mixture, which complexes with the ions, which are formed. After removal of the native oxide, the desired oxide can be grown. This growth is useful because it provides: chemical protection, conditions suitable for lithography, and passivation. The protection prevents unwanted reactions from occurring and the passivation fills vacancies of bonds on the surface not present within the interior of the crystal. Thus, the oxidation of the surface of silicon fulfills several functions in one step.

Thermal oxidation

The growth of oxides on a silicon surface can be a particularly tedious process, since the growth must be uniform and pure. The thickness wanted usually falls in the range 50 - 500 Å, which can take a long time and must be done on a large scale. This is done by stacking the silicon wafers in a horizontal quartz tube while the oxygen source flows over the wafers, which are situated vertically in a slotted paddle (boat), see Figure 8.2. This procedure is performed at 1 atm pressure, and the temperature ranges from 700 to 1200 °C, being held to within ±1 °C to ensure uniformity. The choice of oxidation technique depends on the thickness and oxide properties required. Oxides that are relatively thin and those that require low charge at the interface are typically grown in dry oxygen. When thick oxides are required (>0.5 mm) are desired, steam is the source of choice. Steam can be used at wide range of pressures (1-25 atm), and the higher pressures allow thick oxide growth to be achieved at moderate temperatures in reasonable amounts of time. The thickness of SiO_2 layers on a Si substrate is readily determined by the color of the film. Table 8.1 provides a guideline for thermal grown oxides.

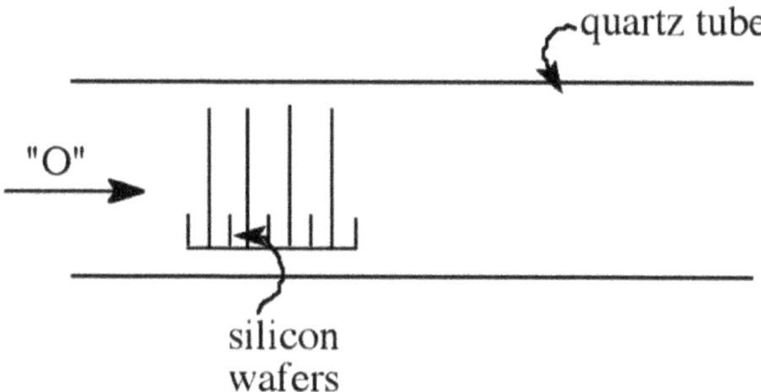

Figure 8.2: Horizontal diffusion tube showing the oxidation of silicon wafers at 1 atm pressure.

Film thickness (μm)	Color	Film thickness (μm)	Color
0.05	Tan	0.63	Violet-red
0.07	Brown	0.68	"Bluish"
0.10	Dark to red-violet	0.72	Blue-green to green
0.12	Royal blue	0.77	"Yellowish"
0.15	Light to metallic blue	0.80	Orange
0.17	Light yellow- green	0.82	Salmon
0.20	Light gold	0.85	Light red-violet
0.22	Gold	0.86	Violet
0.25	Orange to melon	0.87	Blue violet
0.27	Red-violet	0.89	Blue
0.30	Blue to violet blue	0.92	Blue-green
0.31	Blue	0.95	Yellow-green
0.32	Blue to blue-green	0.97	Yellow
0.34	Light green	0.99	Orange
0.35	Green to yellow-green	1.00	Carnation pink
0.36	Yellow-green	1.02	Violet red
0.37	Green-yellow	1.05	Red-violet
0.39	Yellow	1.06	Violet
0.41	Light orange	1.07	Blue-violet
0.42	Carnation pink	1.10	Green
0.44	Violet-red	1.11	Yellow-green
0.46	Red-violet	1.12	Green
0.47	Violet	1.18	Violet
0.48	Blue-violet	1.19	Red-violet
0.49	Blue	1.21	Violet-red
0.50	Blue green	1.24	Carnation pink
0.52	Green	1.25	Orange
0.54	Yellow-green	1.28	"Yellowish"
0.56	Green-yellow	1.32	Sky to green-blue
0.57	"Yellowish"	1.40	Orange
0.58	Light orange to pink	1.46	Blue-violet
0.60	Carnation pink	1.50	Blue

Table 8.1: Color chart for thermally grown SiO$_2$ films observed under daylight fluorescent lighting.

High pressure oxidation

High-pressure oxidation is another method of oxidizing the silicon surface, which controls the rate of oxidation. This is possible because the rate is proportional to the concentration of the oxide, which in turn is proportional to the partial pressure of the oxidizing species, according to Henry's law,

$$C = H_{(pO)}$$

where C is the equilibrium concentration of the oxide, H is Henry's law constant, and p_O is the partial pressure of the oxidizing species. This approach is fast, with a rate of oxidation ranging from 100-1000 mm/h, and also occurs at a relatively low temperature. It is a useful process, preventing dopants from being displaced and also forms a low number of defects, which is most useful at the end of processing.

Plasma oxidation

Plasma oxidation and anodization of silicon is readily accomplished by the use of activated oxygen formed within an electrical discharge or plasma. The oxidation is carried out in a low-pressure (0.05-0.5 Torr) chamber, and the plasma is produced either by a DC electron source or a high-frequency discharge. In simple plasma oxidation the silicon wafer is held at ground potential. In contrast, aniodization systems usually have a DC bias between the sample and an electrode with the sample biased positively with respect to the cathode. Platinum electrodes are commonly used as the cathodes.

There have been at least 34 different reactions reported to occur in an oxygen plasma; however, the vast majority of these are inconsequential with respect to the formation of active species. Furthermore, many of the potentially active species are sufficiently short lived that it is unlikely that they make a significant contribution. The primary active species within the oxygen plasma are undoubtedly O^- and O^{2+}. Both being produced in near equal quantities, although only the former is relevant to plasma aniodization. While these species may be active with respect to surface oxidation, it is more likely that an electron transfer occurs from the semiconductor surface yields activated oxygen species, which are the actual reactants in the oxidation of the silicon.

The significant advantage of plasma processes is that while the electron temperature of the ionized oxygen gas is in excess of 10,000 K, the thermal temperatures required are significantly lower than required for the high-pressure method, i.e., <600 °C. The advantages of the lower reaction temperatures include: the minimization of dopant diffusion and the impediment of the generation of defects. Despite these advantages there are two primary disadvantages of any plasma-based process. First, the high electric fields present during the processes cause damage to the resultant oxide, in particular, a high density of interface traps often results. However, post annealing may improve film quality. Second, the growth rates of plasma oxidation are low, typically 1000 Å/h. This growth rate is increased by about a factor of 10 for plasma aniodization, and further improvements are observed if 1-3% chlorine is added to the oxygen source.

Masking

A selective mask against the diffusion of dopant atoms at high temperatures can be found in a silicon dioxide layer, which can prove to be very useful in integrated circuit processing. A pre-deposition of dopant by ion implantation, chemical diffusion, or spin-on techniques typically results in a dopant source at or near the surface of the oxide. During the initial high-temperature step, diffusion in the oxide must be slow enough with respect to diffusion in the silicon that the dopants do not di use through the oxide in the masked region and reach the silicon surface. The required thickness may be determined by experimentally measuring, at a particular temperature and time, the oxide thickness necessary to prevent the inversion of a lightly doped silicon substrate of opposite conductivity. To this is then added a safety factor, with typical total values ranging from 0.5-0.7 mm. The impurity masking properties of the oxide result it is converted into a silica impurity oxide 'glass' phase, which prevents the impurities from reaching the SiO_2-Si interface.

Applications for silica thin films

While the physical properties of silica make it suitable for use in protective and optical coating applications, the biggest application of insulating SiO_2 thin films is undoubtedly in semiconductor devices, in which the insulator performs a number of specific tasks, including, surface passivation, field effect transistor (FET) gate layer, isolation layers, planarization and packaging.

The term insulator generally refers to a material that exhibits low thermal or electrical conductivity; electrically insulating materials are also called dielectrics. It is in regard to the high resistance to the flow of an electric current that SiO_2 thin films are of the greatest commercial importance. The dielectric constant (ε) is a measure of a dielectric materials ability to store charge and is characterized by the electrostatic energy stored per unit volume across a unit potential gradient. The magnitude of ε is an indication of the degree of polarization or charge displacement within a material. The dielectric constant for air is 1, and for ionic solids is generally in the range of 5 - 10. Dielectric constants are defined as the ratio of the material's capacitance to that of air, i.e.,

$$\varepsilon = C_{material}/C_{air}$$

The dielectric constant for silicon dioxide ranges from 3.9 to 4.9, for thermally and plasma CVD grown films, respectively.

An insulating layer is a film or deposited layer of dielectric material separating or covering conductive layers. Ideally, in these applications an insulating material should have a surface resistivity of greater than 10^{13} Ω/cm^2 or a volume resistivity of greater than 10^{11} $\Omega.cm$. However, for some applications, lower values are acceptable; an electrical insulator is generally accepted to have a resistivity greater than 10^5 $\Omega.cm$. CVD SiO_2 thin films have a resistivity of 10^6-10^{16} $\Omega.cm$, depending on the film growth method.

As a consequence of its dielectric properties SiO_2 and related silicas, are used for isolating conducting layers, to facilitate the diffusion of dopants from doped oxides, as diffusion and ion implantation masks, capping doped films to prevent loss of dopant, for *gettering* impurities, for protection against moisture and oxidation, and for electronic passivation. Of the many methods used for the deposition of thin films, CVD is most often used for semiconductor processing. In order to appreciate the unique problems associated with the CVD of insulating SiO_2 thin films it is worth first reviewing some of their applications. Summarized below are three areas of greatest importance to the fabrication of contemporary semiconductor devices: isolation and gate insulation, passivation, and planarization.

Device isolation and gate insulation

A microcircuit may be described as a collection of devices each consisting of an assembly of active and passive components, interconnected within a monolithic block of semiconducting material. Each device is required to be isolated from adjacent devices in order to allow for maximum efficiency of the overall circuit. Furthermore, within a device, contacts must also be electrically isolated. While there are a number of methods for isolating individual devices within a circuit (reverse-biased junctions, mesa isolation, use of semi-insulating substrates, and oxide isolation), the isolation of the active components in a single device is almost exclusively accomplished by the deposition of an insulator.

In Figure 8.3 is shown a schematic representation of a silicon MOSFET (metal-oxide-semiconductor field effect transistor). The MOSFET is the basic component of silicon-CMOS (complimentary metal-oxide-semiconductor) circuits, which, in turn, form the basis for logic circuits, such as those used in the CPU (central processing unit) of a modern personal computer. It can be seen that the MOSFET is isolated from adjacent devices by a reverse-biased junction (p$^+$-channel stop) and a thick oxide layer. The gate, source and drain contact are electrically isolated from each other by a thin insulating oxide. A similar scheme is used for the isolation of the collector from both the base and the emitter in bipolar transistor devices.

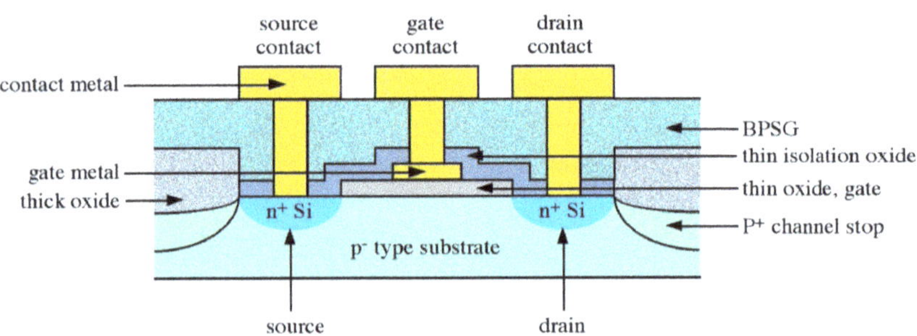

Figure 8.3: Schematic diagrams of a silicon MOSFET (metal-oxide-semiconductor field effect transistor).

As a transistor, a MOSFET has many advantages over alternate designs. The key advantage is low power dissipation resulting from the high impedance of the device. This is a result of the thin insulation layer between the channel (region between source and drain) and the gate contact (Figure 8.3). The presence of an insulating gate is characteristic of a general class of devices called MISFETs (metal-insulator-semiconductor field effect transistor). MOSFETs are a subset of MISFETs where the insulator is specifically an oxide, e.g., in the case of a silicon MISFET device the insulator is SiO_2, hence MOSFET. It is the fabrication of MOSFET circuits that has allowed silicon technology to dominate digital electronics (logic circuits). However, increases in computing power and speed require a constant reduction in device size and increased complexity in device architecture.

Passivation

Passivation is often defined as a process whereby a film is grown on the surface of a semiconductor to:
- chemically protect it from the environment,
- provide electronic stabilization of the surface.

From the earliest days of solid-state electronics, it has been recognized that the presence or absence of surface states plays a decisive role in the usefulness of any semiconducting material. On the surface of any solid-state material there are sites in which the coordination environment of the atoms is incomplete. These sites, commonly termed *dangling bonds*, are the cause of the electronically active states, which allow for the recombination of holes and electrons. This recombination occurs at energies below the bulk value and interferes with the inherent properties of the semiconductor. In order to optimize the properties of a semiconductor device it is desirable to covalently satisfy all these surface bonds, thereby shifting the surface states out of the band gap and into the valence or conduction bands. Electronic passivation may therefore be described as a process, which reduces the density of available electronic states present at the surface of a semiconductor, thereby limiting hole and electron recombination possibilities. In the case of silicon both the native oxide and other oxides admirably fulfill these requirements.

Chemical passivation requires a material that inhibits the diffusion of oxygen, water, or other species to the surface of the underlying semiconductor. In addition, the material is ideally hard and resistant to chemical attack. A perfect passivation material would satisfy both electronic and chemical passivation requirements.

Planarization

For the vast majority of electronic devices, the starting point is a substrate consisting of a flat single crystal wafer of semiconducting material. During processing, which includes the growth of both insulating and conducting films, the surface becomes increasingly non-planar. For example, a gate oxide in a typical MOSFET (see Figure 8.3) may be typically 100-250 Å thick, while the isolation or field oxide may be 10,000 Å. In order for the successful subsequent deposition of conducting layers (metallization) to occur without breaking metal lines (often due to the difficulty in maintaining step coverage), the surface must be at and smooth. This process is called planarization and can be carried out by a technique known as sacrificial etch-back. The steps for the etch-back process are outlined in Figure 8.4.

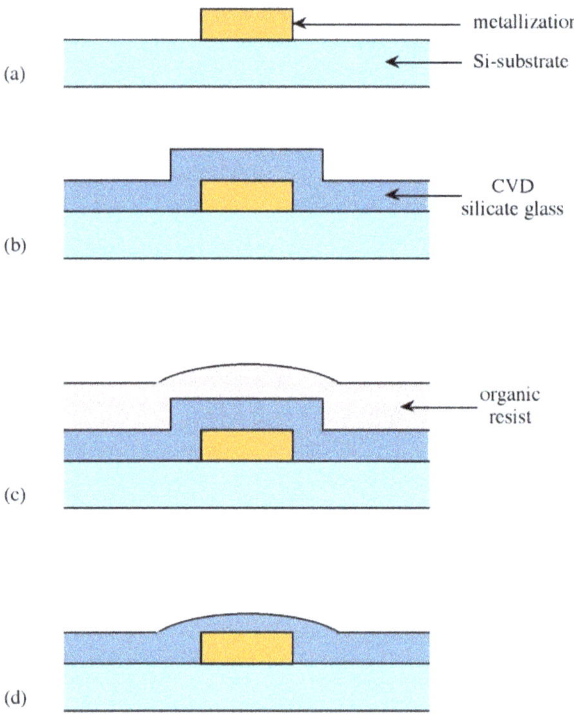

Figure 8.4: Schematic representation of the planarization process. A metallization feature (a) is CVD covered with silicate glass (b), and subsequently coated with an organic resin (c). After etching the resist, a smooth silicate surface is produced (d).

An abrupt step (Figure 8.4a) is coated with a conformal layer of a low melting dielectric, e.g., borophosphorosilicate glass, BPSG (Figure 8.4b), and subsequently a sacrificial organic resin (Figure 8.4c). The sample is then plasma etched such that the resin and dielectric are removed at the same rate. Since the plasma etch follows the contour of the organic resin, a smooth surface is left behind (Figure 8.4d). The planarization process thus reduces step height differentials significantly. In addition, regions or valleys between individual metallization elements (vias) can be completely filled allowing for a route to producing uniformly at surfaces, e.g., the BPSG film shown in Figure 8.4.

The process of planarization is vital for the development of multilevel structures in VLSI circuits. To minimize interconnection resistance and conserve chip area, multilevel metallization schemes are being developed in which the interconnect run in 3-dimensions.

142

Chapter 9: Photolithography

Photolithography, also called optical lithography or UV lithography, is a process used in device micro-fabrication that allows for the creation of pattern regions on a thin film or a wafer. In particular, it is used to differentiate both the structure and chemistry of a specific physical area. For example, to isolate individual devices on a wafer it is necessary to create physical vias (trenches) between them, and etch chemistries are not specific enough to etch one region but not another unless there is a 'barrier' that precludes reaction. In a similar manner, in order to create the channel for a transistor it is necessary to dope a region of the device without affecting the surrounding area. To do this we need to come up with some kind of 'window' to only permit the implanting dopants (impurities) into the semiconductor (Si or GaAs) wafer where we want them and not elsewhere: this is done by constructing an implantation 'barrier'. The creation of these barriers is the basis of photolithography.

Photolithography uses light to transfer a geometric pattern from a photo-mask (also called an *optical mask*) to a photosensitive (that is, light-sensitive) chemical photoresist on the substrate. A series of chemical treatments then either etches the exposure pattern into the material or enables deposition of a new material in the desired pattern upon the material underneath the photoresist.

The first step is generally to grow a layer of silicon dioxide over the entire surface of the wafer. You can grow oxide in either a dry oxygen atmosphere, or in an atmosphere, which contains water vapor, or steam. Figure 9.1 shows the oxide thickness as a function of time for growth with steam: dry O_2 results in a slightly slower growth rate.

On top of the oxide, silicon nitride (Si_3N_4) is deposited by chemical vapor deposition (CVD) by the reaction of dichlorosilane (Figure 4.5) and ammonia in a hot walled low-pressure chemical vapor deposition system (LPCVD). The reaction is:

$$3\,SiH_2Cl_2 + 10\,NH_3 \;\rightarrow\; Si_3N_4 + 6\,NH_4Cl + 6\,H_2$$

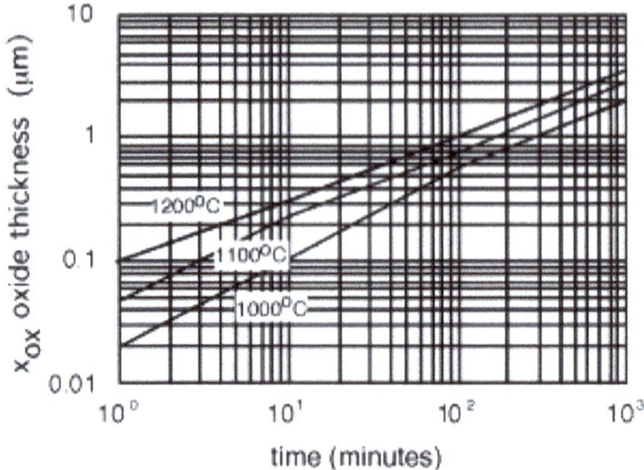

Figure 9.1: A plot of oxide thickness as a function of time for steam oxidation of silicon.

Silicon nitride is a good barrier for oxygen and other impurities, which are not wanted to get into the wafer. At this point the wafer configuration is as shown in Figure 9.2. The silicon wafer is about 250 μm thick (about 0.01") since it has to be strong enough not to break as it is being handled. The two deposited layers are each about 1 μm thick.

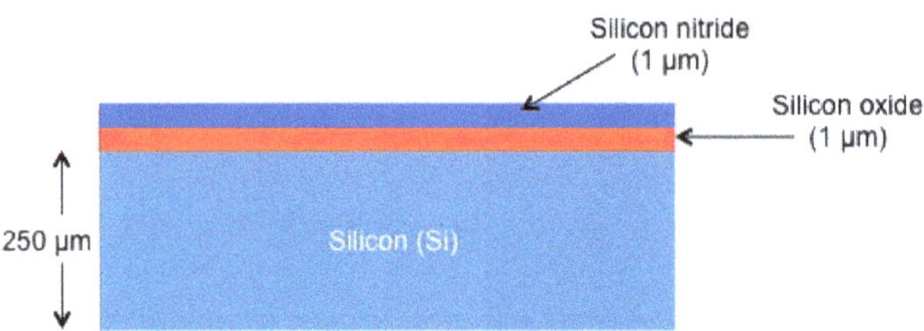

Figure 9.2: Initial wafer configuration. Note the layer thicknesses are not to scale.

In order to selectively remove part of the nitride layer it is necessary to re-move part of the nitride layer using a combination of *photolithography* and *etching*. Firstly, the wafer is coated with a layer of a liquid called photoresist applied through spin coating. The wafer is put on a vacuum chuck, and a layer of liquid photoresist is sprayed on top of the wafer. The chuck is then spun rapidly (at several thousand rpm) such that the centrifugal force causes the resist to spread out uniformly across the wafer surface. The solvent for the photoresist is quite volatile and so the layer of photoresist dries while the wafer is still spinning, resulting in a thin, uniform coating across the wafer Figure 9.3.

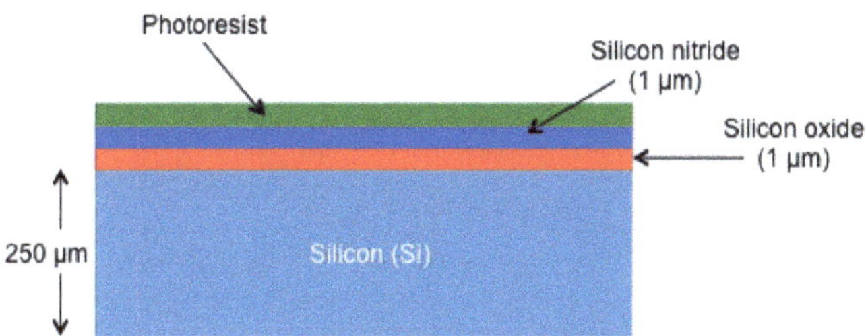

Figure 9.3: The wafer configuration after the photoresist is spin coated. Note the layer thicknesses are not to scale.

A photoresist is a polymer mixed with some kind of light sensitizing compound. With a positive photoresist, wherever light strikes it, the polymer is degraded, and it can be more easily removed with a solvent during the development process. Conversely, negative photoresist is cross-linked when it is illuminated with light, and is more resistant to the solvent than is the unilluminated photoresist. Positive resist is so-called because the image of the developed photoresist on the wafer looks just like the mask that was used to create it. Negative photoresist makes an image, which is the opposite of what the mask looks like.

The instrument that projects the light onto the photoresist on the wafer is called a projection printer or stepper Figure 9.4. As shown in Figure 9.4, the stepper consists of several parts. There is a light source (usually a mercury vapor lamp or ultra-violet excimer lasers), a condenser lens to image the light source on the mask or reticle. The mask contains an image of the pattern we are trying the place on the wafer. The projection lens then makes a reduced (e.g., 5x) image of the mask on the wafer. Because it would be far too costly, if not just plain impossible, to project onto the whole wafer all at once, only a small selected area is printed at one time. Then the wafer is scanned or stepped into a new position, and the image is printed again. If previous patterns have already been formed on the wafer, TV cameras, with artificial intelligence algorithms are used to align the current image with the previously formed features. The stepper moves the whole surface of the wafer under the lens, until the wafer is completely covered with the desired pattern. A stepper is one of the most important pieces of equipment in the whole IC fab; however, since it determines what the minimum feature size on the circuit will be.

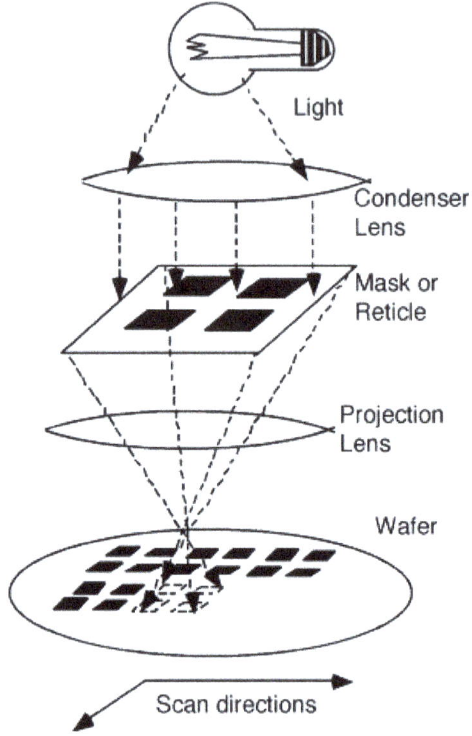

Figure 9.4: A schematic of a stepper configuration.

After exposure, the photoresist is placed in a suitable solvent, and 'developed' as shown in Figure 9.5.

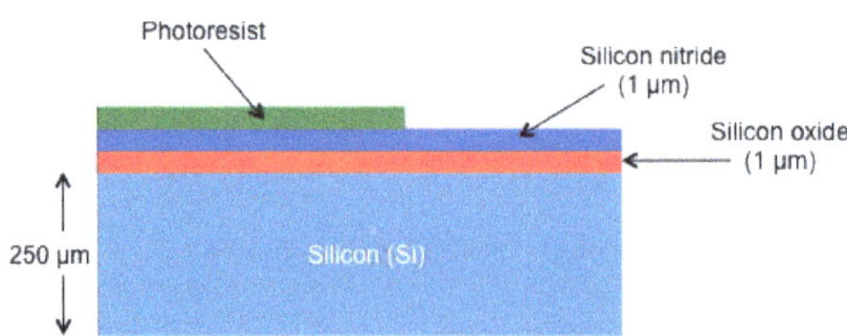

Figure 9.5: The wafer configuration after photoresist exposure and development. Note the layer thicknesses are not to scale.

The pattern that was used in the photolithographic (PL) step exposed half of our area to light, and so the photoresist (PR) in that region was removed upon development. The wafer is now immersed in a hydrofluoric acid (HF) solution, which rapidly etches silicon nitride, but does not etch silicon dioxide nearly as fast, so after the etch we have what we see in Figure 9.6.

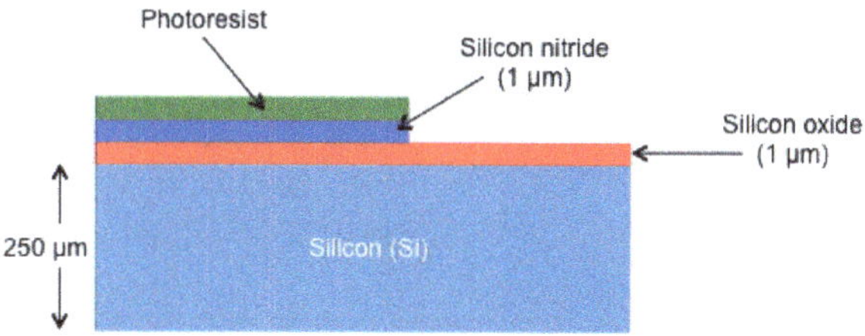

Figure 9.6: The wafer configuration after the nitride etch step. Note the layer thicknesses are not to scale.

The wafer is then placed in the ion implanter and subjected to a 'blast' of phosphorus ions (Figure 9.7). The ions go through the oxide layer on the RHS, but stick in the resist/nitride layer on the LHS of our structure.

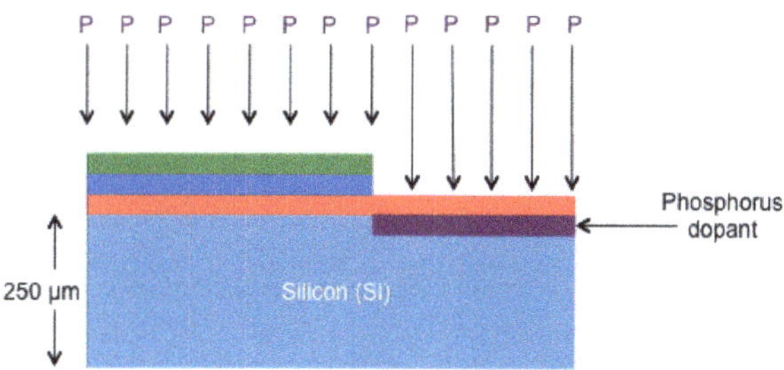

Figure 9.7: The wafer configuration during the implanting phosphorus step. Note the layer thicknesses are not to scale.

Optical issues in photolithography

Photolithography is one of the most important technologies in the production of advanced integrated circuits. It is through photolithography that semiconductor surfaces are patterned and the circuits formed. In order to make extremely small features, on the order of the wavelength of the light, advanced optical techniques are used to transfer a pattern from a mask onto the surface. A polymeric film or *resist* is modified by the light and records the information in a process not dissimilar to ordinary photography.

An illustration of the photolithographic process is shown in Figure 9.8. The process follows the following basic steps:

- The wafer is spin coated with resist to form a uniform ~1 μm thin film of resist on the surface.
- The wafer is exposed with ultraviolet light through a mask, which contains the desired pattern. In the simplest processes the mask is simply placed over the wafer, but advanced sub-micron technologies require the pattern to imaged through a complex optical system.
- The photoresist is developed and the irradiated area is washed away (positive resist) or the un-irradiated area is washed away (negative resist).
- Processing (etching, deposition etc.).
- Remaining resist is stripped.

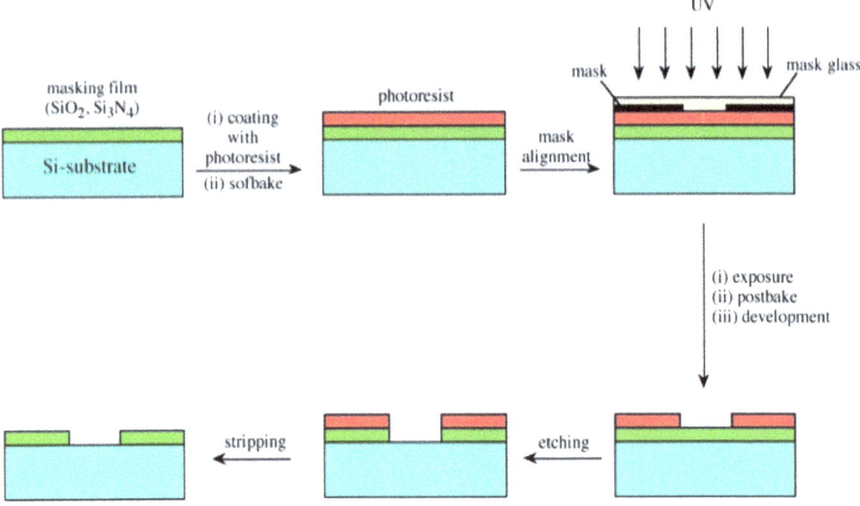

Figure 9.8: Steps in optical printing using photolithography.

In addition to being possibly the most important semiconductor process step, photolithography is also the most expensive technology in semiconductor manufacturing. This expense is the result of two considerations:

1. The optics in photolithography tools is expensive where a single lens can cost a $1 million or more.
2. Each chip (often referred to as a *dye*) must be exposed individually unlike other semiconductor processes such as CVD where an entire wafer can be processed at a time or oxidation processes where many wafers can be processed simultaneously.

This means that not only are photolithography machines the most expensive of semiconductor processing equipment, but more of them are needed in order to maintain throughput.

The critical dimension and depth of focus

Semiconductor process technology is often described by a characteristic length known as the *critical dimension* (CD), which id defined as the smallest feature that needs to be patterned on the surface. The exact definition varies from process to process but is often the channel length of the smallest transistor (typical of a memory chip) or the width of the smallest metal interconnection line (logic chips). This critical dimension is defined by the photolithographic process and is perhaps the most important figure of merit in the manufacture of integrated circuits. Making the critical dimension smaller is the primary focus of improving semiconductor technology for the following reasons:

1. Making the CD smaller dramatically increases the number of devices per unit area and this increase goes with the square of the CD (i.e., a reduction in CD by a factor of 2 generates 4 times the number of devices).
2. Making the CD smaller of a device already in production will make a smaller chip. This means that the number of chips per wafer increases dramatically, and since costs generally scale with the number of wafers and not the number of chips to a wafer, costs are dramatically reduced.
3. Smaller devices are faster.

Therefore, improvements in lithography technology translate directly into better, faster, more complex circuits at lower cost.

Having established the importance of the critical dimension it is important to understand what features of a photolithography system impact. The theory behind projection lithography is very well known, dating from the original analysis of the microscope by Abbe. It is, in fact, the Abbe sine condition that dictates the critical dimension:

$$CD_{Coherent} = 0.82 \frac{\lambda}{n\sin(\theta)}$$

$$CD_{Incoherent} = 0.61 \frac{\lambda}{n\sin(\theta)}$$

where the two expressions refer to the limit of a purely coherent illuminating source and purely incoherent source respectively, and λ is the vacuum wavelength of the illuminating light source, n the index of refraction of the objective lens, and θ refers to the angle between the axis of the lens and the line from the back focal point to the aperture of the entrance of the lens. The quantity in the denominator, $n\sin(\theta)$ is referred to as the numerical aperture (NA). As the degree of coherence can be adjusted in a lithography system, the critical dimension is usually written more generally as:

$$CD = k_1 \frac{\lambda}{n\sin(\theta)}$$

From this equation, we begin to see what can be done to reduce the critical dimension of a lithography system:
- change the wavelength of the source,
- increase the numerical aperture (NA),
- reduce k_1.

The depth of focus (DOF) is the length along the axis in which a sharp image exists. Naturally a large DOF is desirable for ease of alignment, since the entire dye must lie within this region. In reality, however, the more meaningful constraint is that the DOF must be thicker than the resist layer so that the entire volume of resist is exposed and can be developed. Also, if the surface morphology of the device dictates that the resist to be exposed is not planar then the DOF must be large enough so that all features are properly illuminated. Current resists must be 1 µm in thickness in order to have the necessary etch resistance, so this can be considered a minimum value for an acceptable DOF. The depth of focus can also be expressed as a function of numerical aperture and wavelength:

$$DOF = k_2 \frac{\lambda}{[n\sin(\theta)]^2}$$

If we desire to minimize the critical dimension simply by making optics of large numerical aperture that we will simultaneously reduce the DOF and at a much faster rate owing to the dependence on the square of the numerical aperture.

These two quantities, DOF and CD, provide the direction in lithography and semiconductor processing as a whole. For example, a design with an improved surface planarity or a new resist that is effective at smaller thicknesses would allow for a smaller depth of focus, which would in turn allow for a larger numerical aperture implying a smaller critical dimension. The resist, the source wavelength, and the optical delivery system all affect the critical dimension and that further refinements require a multifaceted approach to improving lithography systems. What also must be realized is that, as far as the optical system is concerned, virtually all that can be done with conventional optics has been done and that fundamental restraints on k_1 have been reached.

Wavefront engineering

One way to get around the fundamental limitations of an imaging system is through one of a variety of techniques often termed *wavefront engineering*. Here, not only is the amplitude mapped from the object plane to the image plane, but the phase structure of the light going through the mask is manipulated to improve the contrast and allow for effective values of k_1 lower than the theoretical minimum for uniform illumination. The most important example of these techniques is the phase shift mask or PSM. Here the mask consists of two types of areas, those that allow light to pass through unaffected and some regions where the amplitude of the light is unaffected but its phase is shifted. The resulting electric fields will then sum to zero in some places where use of an ordinary mask would have resulted in a positive intensity.

There are many problems with the practical introduction of various phase shifting techniques. Construction of masks with phase shifting elements, usually a thin layer of poly(methyl methacrylate) (PMMA, Figure 9.9), is difficult and expensive. Mask damage, already a key problem in conventional production techniques, becomes an even greater issue, as traditional mask repair techniques can no longer be used. Also identifying errors in a mask is made more difficult by the odd design.

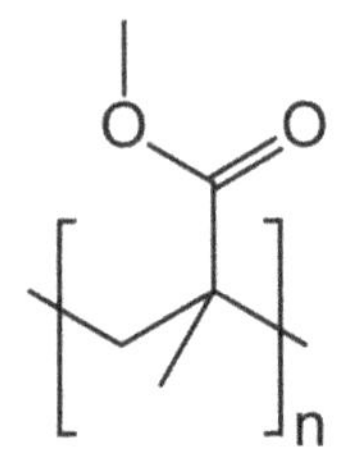

Figure 9.9: Structure of poly(methyl methacrylate) (PMMA).

Interaction with resists

The ultimate resolution of a photolithographic process is not dependent on optics alone, but also on the interaction with the resist. One of the key concerns, particularly as the wavelengths of sources become shorter, is the ability of the source light to penetrate the resist film. Many polymers absorb strongly in the UV, which can limit the interaction to the surface. In such a case only a thin layer of the polymer is exposed and the pattern may not be fully uncovered during developing. One important property of resist is the presence of *saturable absorption.* Saturable absorbers are those absorption sites in the polymer that when excited to a higher state remain there for relatively long periods of time and do not continue to absorb into higher states. If only saturable absorption is present in a polymer film, then continued irradiation eventually leads to transparency, as all absorption sites will be saturated. This allows light penetration through the resist film with full exposure to the substrate surface.

Full penetration of the film leads to a second problem, multiple reflection interference. This occurs when light, which has penetrated the film to the substrate, is then reflected back towards the surface. The result is a standing wave interference pattern, which causes uneven exposure through the film. The problem becomes more severe as optical limits are approached where feature size is approximately equal to the wavelength of the light source meaning such standing waves are the same size as the irradiated features. In the most advanced lithography techniques such as 248 nm lithography with excimer lasers, a special anti-reflectance coating must be laid down before the resist is deposited. Development of an AR coating that has no adverse effects during the exposure and development process is difficult.

One completely new approach to photolithography resists is top-surface-imaged resists or TSI resists. These processes do not require light penetration through the whole volume of resist. In a TSI resist, a silyl amine is selectively diffused from the gas phase into a phenolic polymer in response to the laser irradiation. This diffusion process creates a silyl ether, and development takes place in the form of an oxygen plasma etch, sometimes termed *dry developing*. Depth of focus limitations is avoided as exposure is necessary only at the surface of the resist layer, and the resolution of the etching process determines the final resist profile. Such a technique has tremendous advantages, particularly as source wavelengths become shorter and transparent polymers more rare. Such as resist has a clear optical advantage as well since the image need only be formed at the surface of the resist layer reducing the DOF needed to 100 nm or less, allowing for larger numerical aperture lithography systems with smaller critical dimensions.

Light sources

Current photolithography techniques in production utilize ultraviolet lamps as the light source. In the most production facilities, 0.35 μm mercury i-line technology is used. For the next generation of chips such as 64 Mbit DRAMS better performance is necessary and either i-line technology combined with PSM or a new light source is required. Certainly for the 256 Mbit generation using 0.25 μm technology the i-line source is no longer adequate. The apparent successor is the 248 nm KrF laser, which entered the most advanced production facilities in the late 1990s. KrF technology is often referred to in the literature as Deep UV or DUV lithography. For further shrinkage to 0.18 μm technology, the ArF excimer laser at 193 nm is required.

At critical dimensions lower than 0.18-0.1 μm and below, a whole host of technological problems will need to be overcome in every stage of manufacturing including photolithography. One likely scheme for future lithography is to use X-rays where the wavelength of the light is so much smaller than the feature size such that proximity printing can be used. This is where the mask is placed close to the surface and an X-ray source is scanned across using no optics. Common X-ray sources for such techniques include synchrotron radiation and laser produced plasmas. It has also been widely suggested that the cost of implementing X-ray or other post-optical techniques together with the increased cost of every other manufacturing process step will make improvements beyond 0.1 μm cost prohibitive where benefits in increased circuit speed and density will be dwarfed by massive manufacturing cost. It is noted however that such predictions have been made in the past with regard to other technological barriers.

Composition and photochemical mechanisms of photoresists

In photolithography, a pattern may be transferred onto a photoresist film by exposing the photoresist to light through a mask of the pattern. In the semiconductor industry, the photolithographic procedure includes the following steps as illustrated in Figure 9.9; coating a base material with photoresist, exposing the resist through a mask to light, developing the resist, etching the exposed areas of the base, and stripping the remaining resist off.

Upon exposure to light, the photoresist may become more or less soluble depending on the chemical properties of the particular resist material. The photochemical reactions include chain scission, cross-linking, and the rearrangement of molecules. If the exposed areas of the photoresist become more soluble, then it is a positive resist; conversely, if the exposed resist becomes less soluble, then it is a negative resist. In developing the photoresist, the more soluble material is removed leaving a positive or a negative image of the mask pattern.

Photoresist

Photoresists were initially developed for the printing industry. In the 1920s, the application of photoresists spread to the printed circuit board industry. Photoresists for semiconductor use were first developed in the 1950s; Kodak developed commercial negative photoresists and shortly after, Shipley developed a line of positive resists. Several other companies have entered the market since that time in hopes of manufacturing resist products which meet the increasing demands of the semiconductor industry: narrower line widths, fewer defects, and higher production rates.

Photoresist composition

Several functional requirements must be met for a photoresist to be used in the semiconductor industry. Photoresist polymers must be soluble for easy deposition onto a substrate by spin coating. Good photoresist-substrate adhesion properties are required to minimize undercutting, to maintain edge acuity, and to control the feature sizes. The photoresist must be chemically resistant to whichever etchants are to be used. Sensitivity of the photoresist to a particular light source is essential to the functionality of a photoresist. The speed at which chemical changes occur in a photoresist is its contrast. The contrast of a resist is dependent on the molecular weight distribution of the polymers: a broad molecular weight distribution results in a low contrast resist. High contrast resists produce higher resolution images.

The four basic components of a photoresist are the polymer, the solvent, sensitizers, and other additives. The role of the polymer is to either polymerize or photosolubilize when exposed to light. Solvents allow the photoresist to be applied by spin coating. The sensitizers control the photochemical reactions and additives may be used to facilitate processing or to enhance material properties. Photochemical changes to polymers are essential to the functionality of a photoresist. Polymers are composed primarily of carbon, hydrogen, and oxygen-based molecules arranged in a repeated pattern. Negative photoresists are based on polyisoprene polymers (Figure 9.10); negative resist polymers are not chemically bonded to each other, but upon exposure to light, the polymers crosslink, or polymerize. Positive photoresists are formulated from phenol-formaldehyde novolac resins (Figure 9.11); the positive resist polymers are relatively insoluble, but upon exposure to light, the polymers undergo photosolubilization.

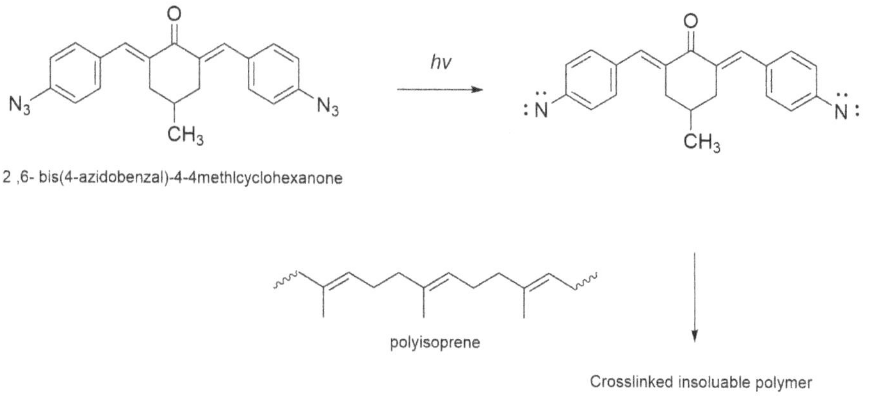

2 ,6- bis(4-azidobenzal)-4-4methlcyclohexanone

polyisoprene

Crosslinked insoluable polymer

Figure 9.10: A crosslinking of a polyisoprene rubber by a photoreactive biazide as negative photoresist.

Solvents are required to make the photoresist a liquid, which allows the resist to be spun onto a substrate. The solvents used in negative photoresists are non-polar organic solvents such as toluene ($C_6H_5CH_3$), xylene [$C_6H_4(CH_3)_2$], and halogenated aliphatic hydrocarbons. In positive resists, a variety of organic solvents such as ethyl cellosolve acetate (2-ethoxyethyl acetate, Figure 9.12), diglyme (Figure 9.13), or cyclohexanone (Figure 9.14) may be used.

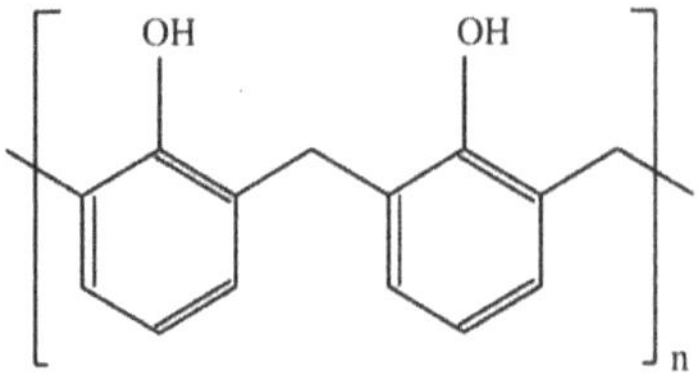

Figure 9.11: The structure of a phenol-formaldehyde (novolac) resin.

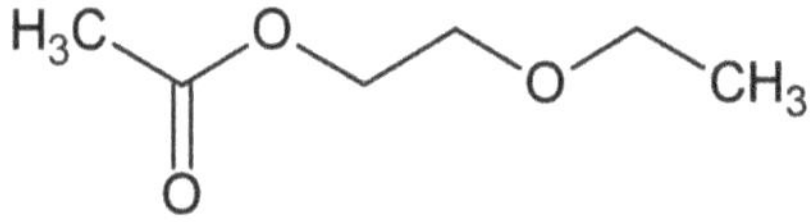

Figure 9.12: Structure of ethyl cellosolve acetate (2-ethoxyethyl acetate).

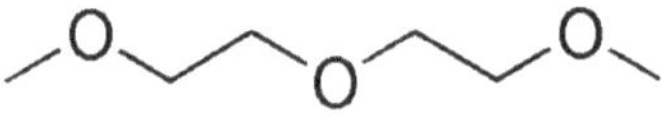

Figure 9.13: Structure of 1-methoxy-2-(2-methoxyethoxy)ethane (diglyme).

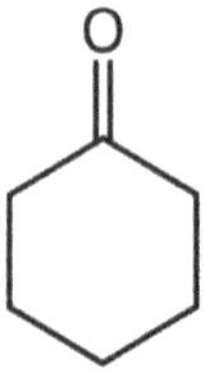

Figure 9.14: Structure of cyclohexanone.

Photosensitizers are used to control or cause polymer reactions resulting in the photosolubilization or crosslinking of the polymer. The sensitizers may also be used to broaden or narrow the wavelength response of the photoresist. Bisazide sensitizers are used in negative photoresists while positive photoresists utilize diazonaphthoquinones (Figure 9.15).

One measure of photosensitizers is their quantum efficiencies, the fraction of photons which result in photochemical reactions; the quantum efficiency of positive diazonaphthoquinone photoresist sensitizers has been measured to be 0.2-0.3 and the quantum efficiency of negative *bis*-arylazide sensitizers is in the range of 0.5-1.0. Additives are also introduced into photoresists depending on the specific needs of the application. Additives may be used to increase photon absorption or to control light within the resist film. Adhesion promoters such as hexamethyldisilazane (Figure 9.16) and additives to improve substrate coating are also commonly used.

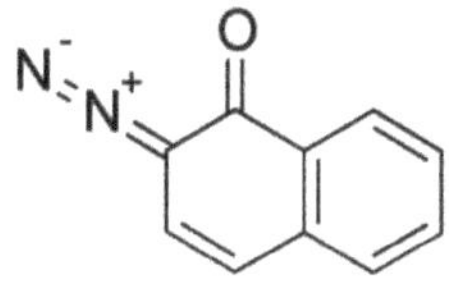

Figure 9.15: Structure of diazonaphthoquinone.

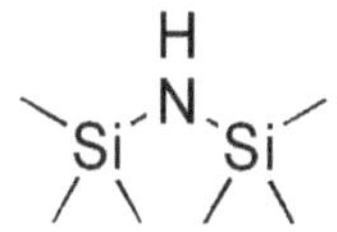

Figure 9.16: Structure of hexamethyldisilazane [1,1,1-trimethyl-*N*-(trimethylsilyl)silanamine].

Negative photoresist chemistry

The matrix resin material used in the formulation of these (negative) resists is a synthetic rubber obtained by a Ziegler-Natta polymerization of isoprene which results in the formation of poly(*cis*-isoprene) (Figure 9.17). Acid-catalyzation of poly(*cis*-isoprene) produces a partially cyclized polymer material; the cyclized polymer has a higher glass transition temperature, better structural properties, and higher density. On the average, microelectronic resist polyisoprenes contain 1-3 rings per cyclic unit, with 5-20% unreacted isoprene units remaining'. The resultant material is extremely soluble in non-polar, organic solvents including toluene, xylene, and halogenated aliphatic hydrocarbons.

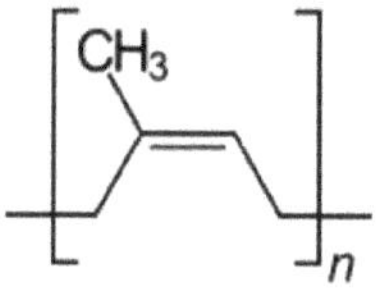

Figure 9.17: Structure of poly(*cis*-isoprene).

The condensation of para-azido benzaldehyde (Figure 9.18) with a substituted cyclohexanone produces *bis*-aryl azide sensitizers. To maximize the absorption of a particular light source the absorbance spectrum of the photoresist may be shifted, by making structural modifications to the sensitizers; for example, by using substituted benzaldehydes, the absorption peak may be shifted to longer wavelengths. A typical *bis*-azide-cyclized polyisoprene photoresist formulation may contain 97 parts cyclized polyisoprene to 3 parts *bis*-azide in a (10 wt%) xylene solvent.

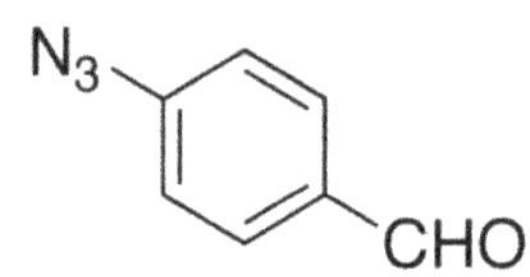

Figure 9.18: Structure of para-azido benzaldehyde.

All negative photoresists function by cross-linking a chemically reactive polymer via a photosensitive agent that initiates the chemical cross-linking reaction. In the bis azide-cyclized polyisoprene resists, the absorption of photons by the photosensitive bis azide in the photoresist results in an insoluble crosslinked polymer. Upon exposure to light, the bis azide sensitizers decompose into nitrogen and highly reactive nitrene intermediate:

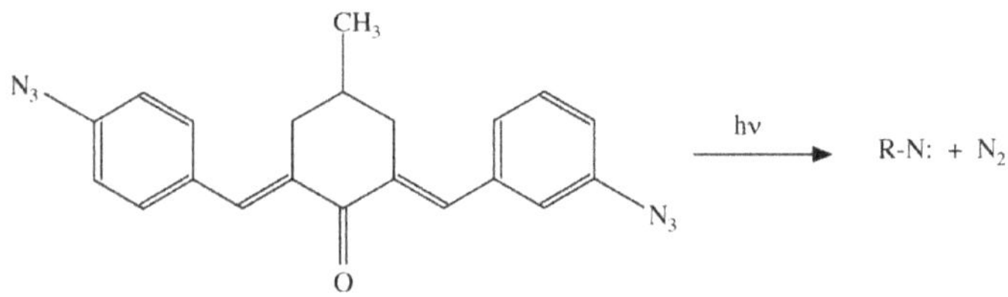

The nitrines react to produce polymer linkages and three-dimensional cross-linked structures that are less soluble in the developer solution.

Positive photoresist chemistry

Positive photoresist materials originally developed for the printing industry have found use in the semiconductor industry. The commonly used novolak resins (phenol-formaldehyde copolymer) and (photosensitive) diazoquinone both were products of the printing industry.

The novolac resin is a copolymer of a phenol and formaldehyde (Figure 9.11). Novolac resins are soluble in common organic solvents (including, ethyl cellosolve acetate and diglyme) and aqueous base solutions. Commercial resists usually contain meta-cresol resins formed by the acid-catalyzed condensation of meta-cresol $[C_6H_4(CH_3)OH]$ and formaldehyde ($H_2C=O$):

The positive photoresist sensitizers are substituted diazonaphthoquinones. The choice of substituents affects the solubility and the absorption characteristics of the sensitizers. Common substituents are aryl sulfonates. The diazoquinones are formed by a reaction of diazonaphthoquinone sulfonyl chloride with an alcohol to form sulfonate ester; the sensitizers are then incorporated into the resist via a carrier or bonded to the resin. The sensitizer acts as a dissolution inhibitor for the novalac resin and is base-insoluble. The positive photoresist is formulated from a novolac resin, a diazonaphthoquinone sensitizer, and additives dissolved in a 20 - 40 wt% organic solvent. In a typical resist, up to 40 wt% of the resist may be the sensitizer.

The photochemical reaction of quinonediazide is:

Base-insoluble sensitizer

Base-soluble photoproduct

Upon absorption of a photon, the quinonediazide decomposes through Wolff rearrangement, specifically a Sus reaction, and produces gaseous nitrogen as a by-product. In the presence of water, the decomposition product forms an indene carboxylic acid, which is base-soluble; however, the formation of acid may not be the reason for increased solubility; the release of nitrogen gas produces a porous structure through which the developer may readily diffuse, resulting in increased solubility.

Image reversal

By introducing an additive to the novolac resins with diazonaphthaquiones sensitizers, the resultant photoresist may be used to form a negative image. A small amount of a basic additive such as imidazole (Figure 9.19) and tri-ethylamine (Figure 9.20) is mixed into a positive novolac resist. Upon exposure to light, the diazonaphthaquiones sensitizer forms an indene car-boxylic acid. During the subsequent baking process, the base catalyzes a thermal decarboxylation, resulting in a substituted indene that is insoluble in aqueous base. Then, the resist is flood exposed destroying the dissolution inhibitors remaining in the previously unexposed regions of the resist. The development of the photoresist in aqueous base results in a negative image of the mask.

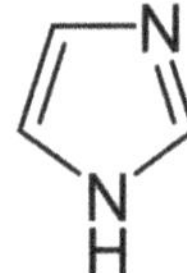

Figure 9.19: Structure of imidazole.

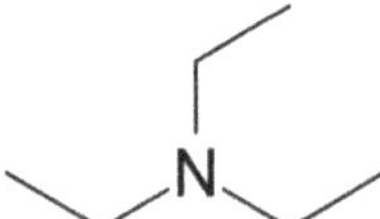

Figure 9.20: Structure of triethylamine.

Comparison of positive and negative photoresists

Up to the 1970s, negative photoresist processes dominated. The poor adhesion and the high cost of positive photoresists prevented its widespread use at the time. As device dimensions grew smaller, the advantages of positive photoresists, better resolution and pinhole protection, suited the changing demands of the semiconductor industry and in the 1980s the positive photoresists came into prominence. A comparison of negative and positive photoresists is given in Figure 9.21.

The better resolution of positive resists over negative resists may be attributed to the swelling and image distortion of negative resists during development; this prevents the formation of sharp vertical walls of negative resist. Disadvantages of positive photoresists include a higher cost and lower sensitivity.

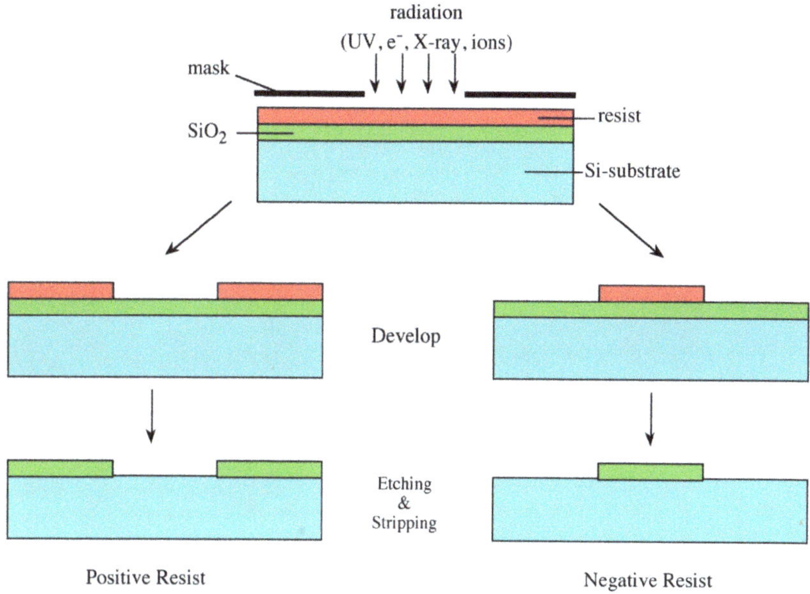

Figure 9.21: A comparison of negative and positive photoresists.

Positive photoresists have become the industry choice over negative photoresists. Negative photoresists have much poorer resolution and the positive photoresists exhibit better etch resistance and better thermal stability. As optical masking processes are still preferred in the semiconductor industry, efforts to improve the processes are ongoing. Currently, researchers are studying various forms of chemical amplification to increase the photon absorption of photoresists.

Chapter 10: Thin Film Growth

A thin film is a layer of material ranging from fractions of a nanometer (monolayer) to several micrometers in thickness. The controlled synthesis of materials as thin films is commonly referred to as *deposition* or *growth*, and can be categorized as being either physical or chemical.

Physical deposition uses mechanical, electromechanical or thermodynamic means to produce a thin film of solid. While the elemental sources are ordinarily solids or liquids physical deposition systems tend to require a low-pressure vapor environment to function properly, and thus, most can be classified as physical vapor deposition (PVD).

By contrast chemical deposition involves a molecular precursor undergoing a chemical change (usually decomposition to some extent) at a solid surface, leaving a solid layer. Chemical deposition is further categorized by the phase of the precursor:
- liquid including liquid phase deposition (LPD), chemical solution deposition (CSD), chemical bath deposition (CBD),
- vapor including chemical vapor deposition (CVD) and atomic layer deposition (ALD).

Epitaxy

Epitaxy, is a transliteration of two Greek words *epi*, meaning "upon", and *taxis*, meaning "ordered". With respect to crystal growth it applies to the process of growing thin crystalline layers on a crystal substrate. In epitaxial growth, there is a precise crystal orientation of the film in relation to the substrate. The growth of epitaxial films can be achieved by a number of methods including molecular beam epitaxy (MBE), atomic layer epitaxy (ALE), and chemical vapor deposition (CVD), all of which will be described later.

Epitaxy of the same material, such as a gallium arsenide film on a gallium arsenide substrate, is called *homoepitaxy*, while epitaxy where the film and substrate material are different is called *heteroepitaxy*. Clearly, in homoepitaxy, the substrate and film will have the identical structure, however, in heteroepitaxy, it is important to employ where possible a substrate with the same structure and similar lattice parameters. For example, zinc selenide (zinc blende, a = 5.668 Å) is readily grown on gallium arsenide (zinc blende, a = 5.653 Å). Alternatively, epitaxial crystal growth can occur where there exists a simple relationship between the structures of the substrate and crystal layer, such as is observed between γ-Al$_2$O$_3$ (100) on Si (100), see Figure 10.1.

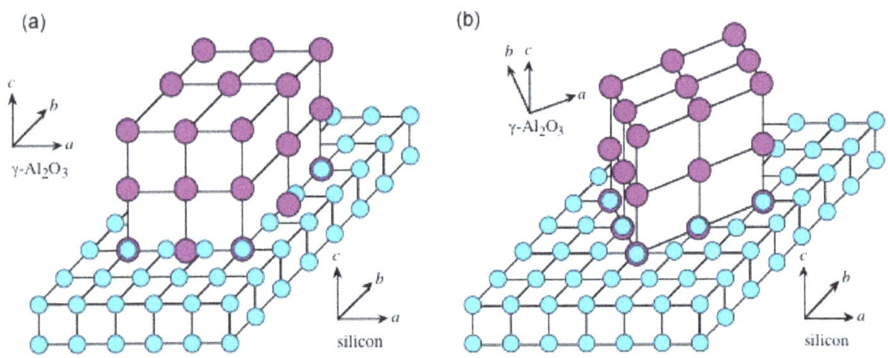

Figure 10.1: Schematic diagram of the two different lattice-matching crystallographic relationships between γ-Al$_2$O$_3$ on Si(100): (a) γ-Al$_2$O$_3$ (100)‖Si(100), and (b) γ-Al$_2$O$_3$ (100)‖Si(110). Adapted from K. Sawada, M. Ishida, T. Nakamura, and N. Ohtake, Metalorganic molecular beam epitaxy of γ-Al$_2$O$_3$ films on Si at low growth temperatures. *Appl. Phys. Lett.*, **1988, 52, 1673. Copyright: American Institute of Physics (1988).**

Whichever route is chosen a close match in the lattice parameters is required, otherwise, the strains induced by the lattice mismatch results in distortion of the film and formation of dislocations. If the mismatch is significant epitaxial growth is not energetically favorable, causing a textured film or polycrystalline un-textured film to be grown. As a general rule of thumb, epitaxy can be achieved if the lattice parameters of the two materials are within about 5% of each other. For good quality epitaxy, this should be less than 1%. The larger the mismatch, the larger the strain in the film, since as the film gets thicker, it will try to relieve the strain, which can result in the loss of epitaxy through the formation of a dislocation (Figure 10.2). It is important to note that the <100> directions of a film must be parallel to the <100> direction of the substrate. In some cases, such as Fe on MgO, the [111] direction is parallel to the substrate [100]. The epitaxial relationship is specified by giving first the plane in the film that is parallel to the substrate [100].

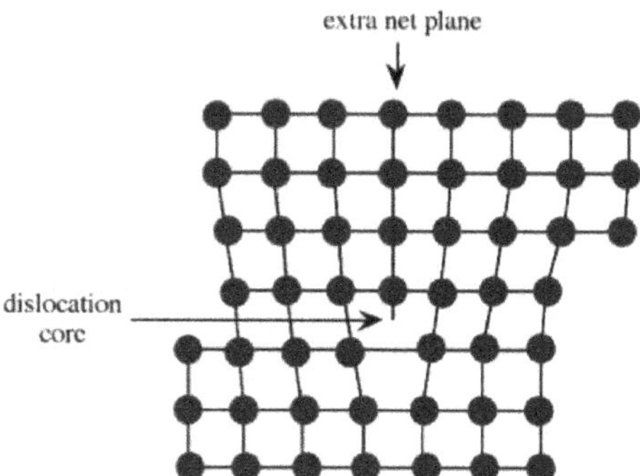

Figure 10.2: Schematic representation of a dislocation in a crystal lattice.

Sputtering

Sputter deposition is a physical vapor deposition (PVD) method of thin film deposition that involves ejecting material from a *target* that is a source onto a substrate such as a silicon wafer.

The most common application of the sputter system is in metallization, in which the entire surface of the substrate is coated with a conductor; however, masking or lithography allows for specific structures to be created. In most MOS devices an aluminum-silicon alloy is usually used, although other metals are employed as well, e.g., front contacts in crystalline silicon solar cells are silver.

A sputtering system is shown schematically in Figure 10.3. A sputtering system is a vacuum chamber, which after it is pumped out, is re-filled with a low-pressure argon gas. A high voltage ionizes the gas, and creates what is known as the Crookes dark space near the cathode, which in our case, consists of a metal target made out of the metal we want to deposit. Almost all of the potential of the high-voltage supply appears across the dark space. The glow discharge consists of argon ions and electrons, which have been stripped off of them. Since there are about equal number of ions and electrons, the net charge density is about zero, and hence by Gauss' law, so is the field. The electric field accelerates the argon atoms, which slam into the aluminum target. There is an exchange of momentum, and an aluminum atom is ejected from the target (Figure 10.4) and heads to the silicon wafer, where it sticks, and builds up a metal film.

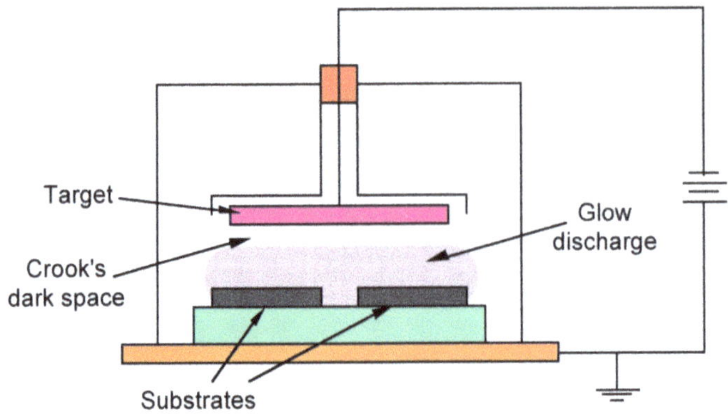

Figure 10.3: A schematic representation of a sputtering apparatus.

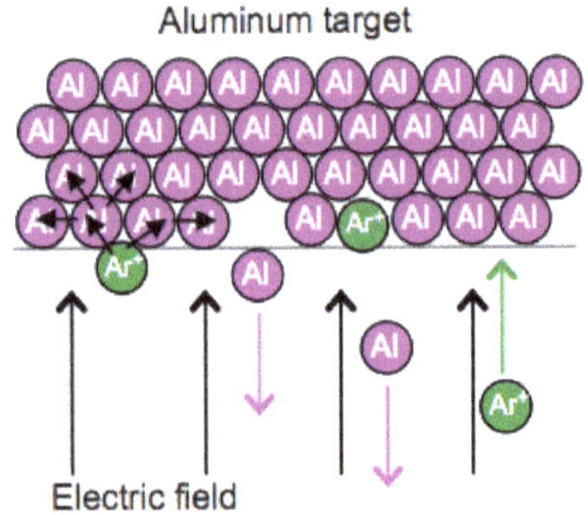

Figure 10.4: The sputtering mechanism.

Molecular beam epitaxy

In the most basic form of molecular beam epitaxy (MBE), the substrate is placed in ultra high vacuum (UHV) and the materials for the film are evaporated from elemental sources. The evaporated molecules or atoms flow as a beam, striking the substrate, where they are adsorbed on the surface. Once on the surface, the atoms move by surface diffusion until they reach a thermodynamically favorable location to bond to the substrate. Molecules will dissociate to their atomic form either during diffusion or once at a favorable site. Figure 10.5 illustrates the processes that can occur on the surface. Because the atoms require time for surface diffusion, the quality of the film will be better with slower growth. Typically growth rates of about 1 monolayer per second provide sufficiently high quality.

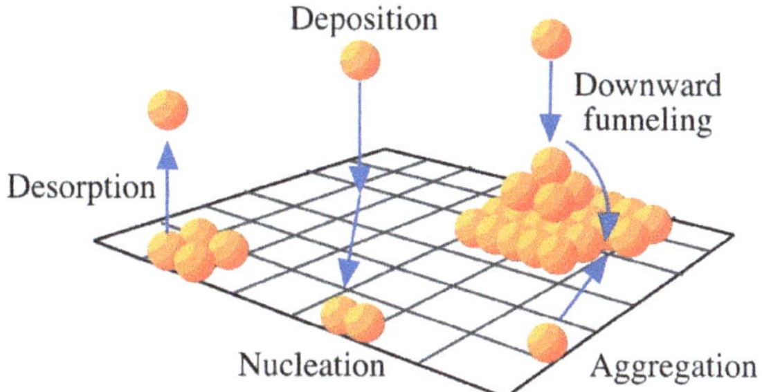

Figure 10.5: Schematic illustration of the processes occurring on the growth surface during MBE: adsorption of atoms on the surface, surface diffusion of atoms, formation of crystalline lattice, and desorption of particles from the surface.

A typical MBE chamber is shown in Figure 10.6. The substrate is chemically washed and then put into a loading chamber where it is further cleaned using argon ion bombardment followed by thermal annealing. This removes the top layers of the substrate, which is usually an undesired oxide that grew in air and contains impurities. The annealing heals any damage caused by the bombardment. The substrate then enters the growth chamber via the sample exchange load lock. It is secured on a molybdenum holder either mechanically or with melted indium or gallium which hold the substrate by surface tension.

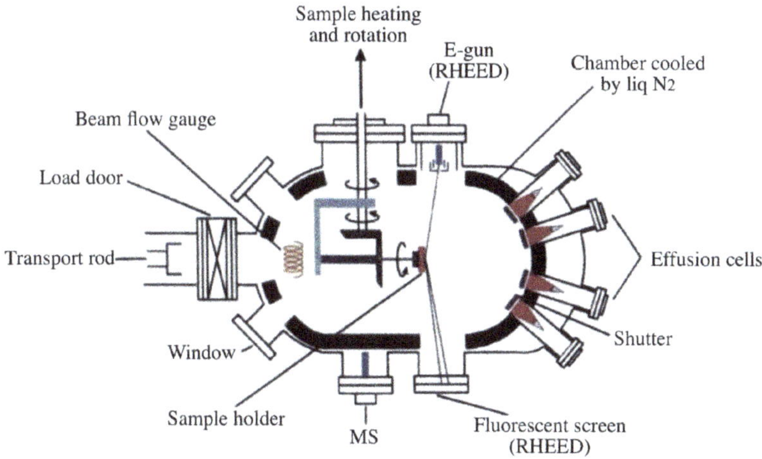

Figure 10.6: Schematic view of a typical MBE growth chamber.

Each effusion cell (Figure 10.6) is a source of one element in the film. The effusion cell, also called a Knudsen cell, contains the elemental form in very high purity: in the case of Ga and As, for GaAs growth, this greater than 99.99999%. The cell is heated to encourage evaporation of the elements. For GaAs growth, the temperature is typically controlled to create a vapor pressure of 10^{-2}-10^{-3} Torr (i.e., 0.13-0.013 Pa) inside the effusion cell, which results in a transport of about 10^{15} molecules/cm^2 to the substrate when the shutter for that cell is opened. The shape and size of the opening in the cell is optimized for an even distribution of particles on the substrate. Due to the relatively low concentration of molecules, they typically do not interact with other molecules in the beam during the 5-30 cm journey to the substrate. The substrate is usually rotated, at a few rpm, to further even the distribution.

Because MBE takes place in UHV and has relatively low pressure of residual gas at the surface, analysis techniques such as reflection high-energy diffraction (RHEED) and ellipsometry can be used during growth, both to study and control the growth process. The UHV environment also allows pre or post growth analysis techniques such as Auger spectroscopy.

Elemental and molecular sources

The effusion cell is used for the majority of MBE growth. All materials used in the cell are carefully chosen to be non-interacting with the element being evaporated. For example, the crucible is commonly pyrolitic boron nitride; however, it has disadvantages, such as:

- the evaporated species may be molecular, rather than monomeric, which will require further dissociation at the surface,
- when the shutter is opened, the heat loss from the cell results in a transient in the beam flux, which last for several minutes and cause variations of up to 50%,
- the growth chamber must be opened up to replace the solid sources.

Cracker cells are used to improve the ratio of monomeric to molecular (or at least dimeric to tetrameric) particles from the source. The cracker cell, placed so that the beam passes through it after the effusion cell, is maintained at a high temperature (and sometimes high pressure) to encourage dissociation. The dissociation process generally requires a catalyst and the best catalysts for a given species have been studied.

Some elements, such as silicon, have low enough vapor pressure that more direct heating techniques such as electron bombardment or laser radiation heating are used. The electron beam is bent using electromagnetic focusing to prevent any impurities in the electron source from contaminating the silicon to be used in MBE. Because the heat is concentrated on the surface to be evaporated, interactions with and contamination from the crucible walls is reduced. In addition, this design does not require a shutter, so there is no problem with transients. Modulation of the beam can produce very sharp interfaces on the substrate. In laser radiation heating, a laser beam replaces the electron beam. The advantages of localized heating and rapid modulation are also maintained without having to worry about contamination from the electron source or stray electrons.

Some of the II-VI (12-16) compounds have such high vapor pressure that a Knudson cell cannot be used. For example, a mercury source must be kept cooler than the substrate to keep the vapor pressure low enough to allow controlled growth. The Hg source must also be sealed off from the growth chamber to allow the chamber to be pumped down.

Two other methods of obtaining the elements for use in epitaxy are gas-source epitaxy and chemical beam epitaxy (CBE). Both of these methods use gas sources, but they are distinguished by the use of elemental beams in gas source epitaxy, while organometallic beams are used in CBE. For the example of III-V (13-15) semiconductors, in gas epitaxy, the group III material may come from an effusion cell, while the Group V material is the hydride, such as AsH_3 or PH_3, which is cracked before entering the growth chamber. In CBE, the group V material is an organometallic compound, such as triethylgallium $[Ga(C_2H_5)_3]$ or trimethylaluminum $[Al(CH_3)_3]$, which adsorbs on the surface, where it dissociates.

Gas sources have several advantages. Gas lines can be run into the chamber, which allows the supply to be replenished without opening the chamber. When making alloys, such as $Al_xGa_{1-x}As$, the gases can be premixed for the correct stoichiometry or even have their composition gradually changed for making graded structures. For abrupt structures, it is necessary to be able to switch the gas lines with speeds of 1 second or less; however, the gas lines increase the complexity of the process and can be hard on the pumping system.

Substrate choice and preparation

Materials can be grown on substrates of different structure, orientation, and chemistry. In deciding which materials can be grown on a particular substrate, a primary consideration was expected to be lattice mismatch, i.e., differences in spacing between atoms. Although lattice mismatch can cause strain in the grown layer, considerable accommodation between materials of different sizes can take place during growth. A greater source of strain can be differences in thermal expansion characteristics because the layer is grown at high temperature. On cooling to room temperature, dislocation defects can be formed at the interface or in severe cases the device may break. Chemical considerations, such as whether the layer's elements will dissolve in the substrate or form compounds with the substrate, must also be considered.

Different orientations of the substrate can also affect growth. Close-packed planes have the lowest surface energy, which allows atoms to desorb from the surface, resulting in slower growth rates. Growth is favored where bonds can be made in several directions at the same time. Therefore, the substrate is often cut off-axis by a 2-4° to provide a rougher growth surface. For compound semiconductors, some orientations result in the number of loose bonds changing between layers. This results in changes of surface energy with each layer, which slows growth down.

The greatest cause of defects in the epitaxial layer is defects on the substrate's surface. In general, any dislocations on the substrate are replicated or even multiplied in the epitaxial growth, which is what makes the cleaning of the substrate so important.

Materials grown

MBE is commercially used primarily for GaAs devices. This is partly because the high-speed microwave devices made from GaAs required the superior electrical quality of epitaxial layers. Taking place at lower temperature and under better-controlled conditions, MBE generally results in layers of better quality than melt-grown.

From solid gallium and arsenic sources, elemental Ga and tetrameric As_4 are evaporated. For a GaAs substrate, the Ga flux has a sticking coefficient very close to 1 (almost certain to adsorb), and in contrast, As is much less likely to adsorb, so an excess of the latter is usually supplied. Cracker cells are often used on the As_4 in order to obtain As_2 instead, which results in faster growth and more efficient utilization of the source beam.

For nominally un-doped GaAs grown by MBE, the residual impurities are usually carbon, from substrate contamination and residual gas after the growth chamber is pumped down, and sulfur, from the As source. The most common surface defects are *oval* defects (e.g., Figure 10.7), which seem to form when Ga manages to form metallic droplets during the growth process, which can particularly occur if the substrate was not cleaned properly. These defects can also be reduced by carefully controlling the Ga flux.

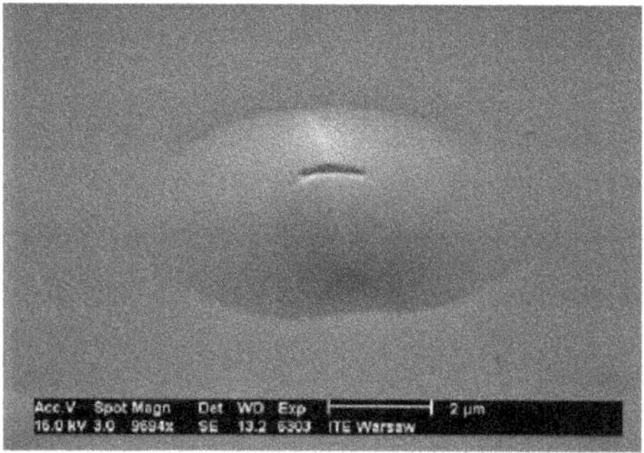

Figure 10.7: SEM image of an example of a Ga cell related oval defects with a core. Adapted from A. Szerling, K. Kosiel, M. Płuska, T. Ochalski, and J. Ratajczak, Oval defects in crystals grown by MBE technique: study and methods of elimination. *Electron Technol.*, **2004, 36, 1. Copyright: Institute of Electron Technology (2004).**

During MBE growth, dopants can be introduced by having a separate effusion cell or gas source for each dopant. To achieve a desired dopant concentration in the film, not only must the rate of dopants striking the substrate be controlled, but, the characteristics of how the dopant behaves on the surface must be known. Low-vapor pressure dopants tend to desorb from the surface and their behavior is very temperature dependent and so are avoided when possible. Slow diffusing dopants adsorb to surface sites and are eventually covered, as more GaAs is grown. Their incorporation depends linearly on the partial pressure of the dopant present in the growth chamber. This is the behavior exhibited by most n-type dopants in GaAs and most dopants of both types in Si. If the dopant, like most p-type GaAs dopants, is able to diffuse through the surface of the substrate into the crystal below, then there will be higher incorporation efficiency, which will depend on the square root of the dopant partial pressure for reasonable concentrations. Due to increasing lattice strain, all dopants will saturate at very high concentrations. They may also tend to form clusters. Dopant behavior depends on many factors and is actively studied.

The growth of GaAs epitaxial layers on silicon substrates has also been investigated. Silicon substrates are grown in larger wafers, have better thermal conductivity allowing more devices/chip to be grown on them, and are cheaper; however, because Si is nonpolar and GaAs is polar, the GaAs tends to form islands on the surface with different phase (what should be a Ga site based on a neighboring domain's pattern will actually be an As site). There is also a fairly large lattice mismatch leading to many dislocations. Nevertheless, FETs, LEDs, and lasers have all been made by MBE.

Many devices require abrupt junctions between layers of different materials. One group, studying how to make high quality, abrupt GaAs and AlAs layers, found that rapid movement of the Ga or Al on the surface was required. This migration was enhanced at high temperatures, but unfortunately, diffusion into the substrate also increased; however, they also discovered that migration of Ga or Al increased if the As supply was turned off. By alternating the Ga and As supplies, the Ga was able to reach the substrate and migrate to provide more even monolayer coverage before the As atoms arrived to react.

Besides GaAs, most other III-V (13-15) semiconductors have also been grown using MBE. Structures involving very thin layers (only a few atomic layers thick), called *superlattices* (or strained superlattices if there is a large lattice mismatch) are routinely grown. Because different materials have different energy levels for electrons and holes, it is possible to trap carriers in one of these thin layers, forming a quantum well. This type of confinement structure is particularly popular for LEDs or lasers, including blue light lasers. The strained superlattice structure actually shifts and splits the energy levels of the materials in some cases making devices possible for such applications as infrared light detection, which requires very small band gaps.

Thin films of many other materials have also been grown using MBE methods. Silicon technology has cheaper methods of growth and so Si layers are not very popular; however, possible devices made of Si-Ge alloys have been grown and II-VI compounds have also been grown. Magnetic materials, such as Co-Pt and Fe-Pt alloys, have been grown in the hopes of providing better magnetic storage.

Analysis techniques

The most popular in-situ analysis technique for MBE-grown layers is reflection high-energy diffraction (RHEED). Electrons of energy 5 - 40 keV are directed towards the sample, where they reflect from the surface at a very small angle (less than 3°) and are directed onto a screen (Figure 10.8). These electrons interact with only the top few atomic layers and thus provide information about the surface.

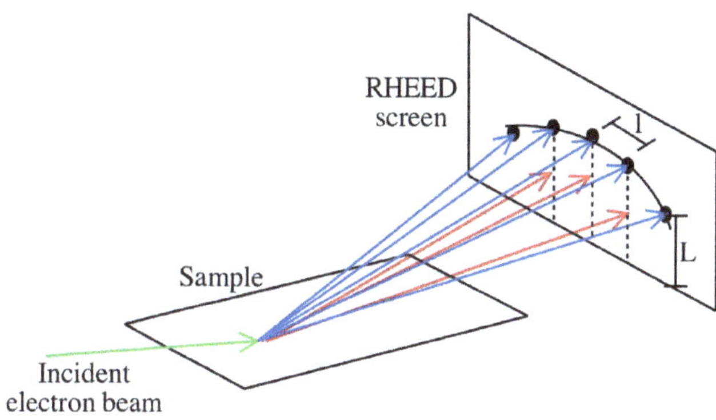

Figure 10.8: Schematic illustrating the formation of a reflection high energy diffraction (RHEED) pattern.

Figure 10.9 shows a typical pattern on the screen for electrons reflected from a smooth surface in which constructive interference between some of the electrons reflected from the lattice structure results in lines. If the surface is rough, spots will appear on the screen, also. By looking at the total intensity of the reflected electron pattern, an idea of the number of monolayers deposited and how epilayers grow can be obtained. The island-type growth shown in this figure is an area of intense interest. These oscillations in intensity are gradually damped as more layers are grown, because the overall roughness of the surface increases.

Phase locked epitaxy takes advantage of the patterns of the oscillations in RHEED to grow very abrupt layers. By sending the oscillation information to a computer, it can decide when to open or close the shutters of the effusion cell based on the location in the oscillation cycle. This technique self-adjusts for fluctuations in beam flux when the shutters are opened and can grow very abrupt layers.

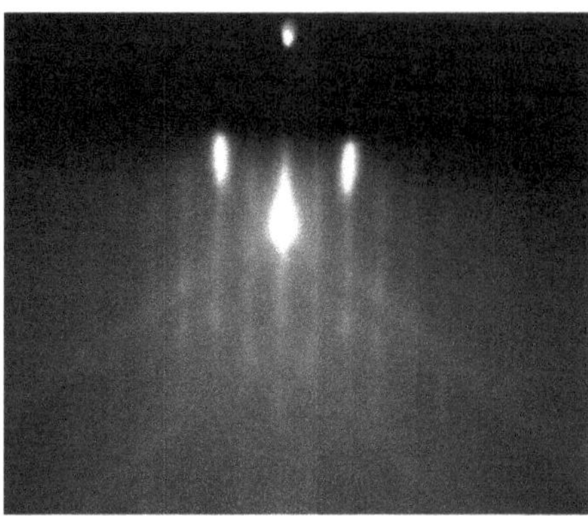

Figure 10.9: RHEED diffraction pattern of a GaAs surface. Adapted from images by the MBE Laboratory in the Institute of Physics of the Czech Academy of Sciences.

Another analysis technique that can be used to study surface smoothness during growth is ellipsometry. Polarized laser light is reflected from the surface at a small angle. The polarization of the light changes, depending on the roughness of the surface.

Improved growth characteristics also require that the actual flux from the sources be measured. This is typically done with an ion gauge flux monitor, which is either used to measure residual beam that misses the substrate or is moved into the beam path for calibration when a new source is used.

Because of the importance of clean substrate surfaces for low-defect growth, Auger spectroscopy is used following cleaning by sputtering. Auger spectroscopy takes place by ionizing an inner shell electron from an atom. When an outer shell electron then de-excites to the inner shell, the energy released can prompt the emission of another outer shell electron. The energy at which this occurs is characteristic of the atom involved and the signal can be used to detect impurities as small as 0.1%.

Atomic layer deposition

Atomic layer deposition (ALD), originally called atomic layer epitaxy (ALE), was first reported by in the 1970's for the growth of zinc sulfide (ZnS) thin films to fabricate electroluminescent flat panel displays.

ALD refers to the method whereby film growth occurs by exposing the substrate to its starting materials alternately. It should be noted that ALE is actually a sub-set of ALD, in which the grown film is epitaxial to the substrate; however, the terms are often used interchangeably. Although both ALD and CVD use chemical (molecular) precursors, the difference between the techniques is that the former uses self limiting chemical reactions to control in a very accurate way the thickness and composition of the film deposited. In this regard ALD can be considered as taking the best of CVD (the use of molecular precursors and atmospheric or low pressure) and MBE (atom-by-atom growth and a high control over film thickness) and combining them in single method. A selection of materials deposited by ALD is given in Table 10.1.

Compound class	Examples
II–VI compounds	ZnS, ZnSe, ZnTe, $ZnS_{1-x}Se_x$, CaS, SrS, BaS, $SrS_{1-x}Se_x$, CdS, CdTe, MnTe, HgTe, $Hg_{1-x}Cd_xTe$, $Cd_{1-x}Mn_xTe$
II–VI based thin-film electroluminescent (TFEL) phosphors	ZnS:M (M = Mn, Tb, Tm), CaS:M (M = Eu, Ce, Tb, Pb), SrS:M (M = Ce, Tb, Pb, Mn, Cu)
III–V compounds	GaAs, AlAs, AlP, InP, GaP, InAs, $Al_xGa_{1-x}As$, $Ga_xIn_{1-x}As$, $Ga_xIn_{1-x}P$
Semiconductors/dielectric nitrides	AlN, GaN, InN, SiN_x
Metallic nitrides	TiN, TaN, Ta_3N_5, NbN, MoN
Dielectric oxides	Al_2O_3, TiO_2, ZrO_2, HfO_2, Ta_2O_5, Nb_2O_5, Y_2O_3, MgO, CeO_2, SiO_2, La_2O_3, $SrTiO_3$, $BaTiO_3$
Transparent conductor oxides	In_2O_3, In_2O_3:Sn, In_2O_3:F, In_2O_3:Zr, SnO_2, SnO_2:Sb, ZnO,
Semiconductor oxides	ZnO:Al, Ga_2O_3, NiO, CoO_x
Superconductor oxides	$YBa_2Cu_3O_{7-x}$
Fluorides	CaF_2, SrF_2, ZnF_2

Table 10.1: Examples of thin film materials deposited by ALD. Adapted from M. Ritala and M. Leskel, Atomic layer epitaxy - a valuable tool for nanotechnology? *Nanotechnology,* **1999, 10, 19.**

How ALD works

The premise behind the ALD process is a simple one. The substrate (amorphous or crystalline) is exposed to the first gaseous precursor molecule (elemental vapor or volatile compound of the element) in excess and the temperature and gas flow is adjusted so that only one monolayer of the reactant is chemisorbed onto the surface (Figure 10.10a).

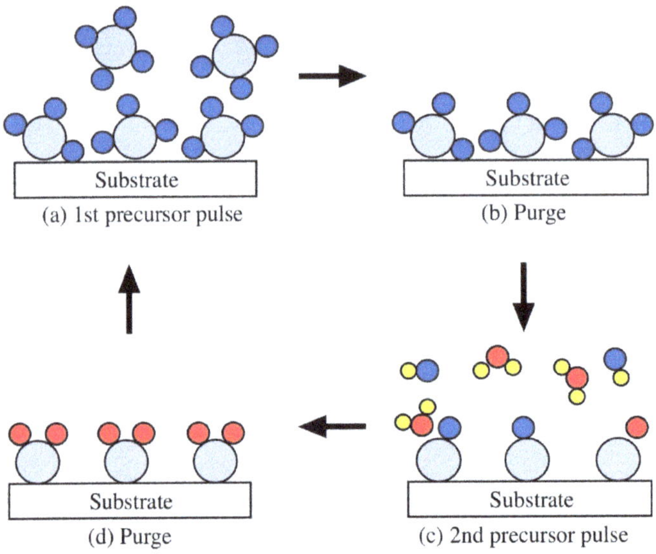

Figure 10.10: Schematic representation of an ALD process.

The excess of the reactant, which is in the gas phase or physisorbed on the surface, is then purged out of the chamber with an inert gas pulse before exposing the substrate to the other reactant (Figure 10.10b). The second reactant then chemisorbs and undergoes an exchange reaction with the first reactant on the substrate surface (Figure 10.10c). This results in the formation of a solid molecular film and a gaseous side product that may then be removed with an inert gas pulse (Figure 10.10d).

The deposition may be defined as self-limiting since one, and only one, monolayer of the reactant species remains on the surface after each exposure. In this case, one complete cycle results in the deposition of one monolayer of the compound on the substrate. Repeating this cycle leads to a controlled layer-by-layer growth. Thus the film thickness is controlled by the number of precursor cycles rather than the deposition time, as is the case for a CVD processes. This self-limiting behavior is the fundamental aspect of ALD and understanding the underlying mechanism is necessary for the future exploitation of ALD.

One basic condition for a successful ALD process is that the binding energy of a monolayer chemisorbed on a surface is higher than the binding energy of subsequent layers on top of the formed layer; the temperature of the reaction controls this. The temperature must be kept low enough to keep the monolayer on the surface until the reaction with the second reactant occurs, but high enough to re-evaporate or break the chemisorption bond. The control of a monolayer can further be influenced with the input of extra energy such as UV irradiation or laser beams. The greater the difference between the bond energy of a monolayer and the bond energies of the subsequent layers, the better the self-controlling characteristics of the process.

Basically, the ALD technique depends on the difference between chemisorption and physisorption. Physisorption involves the weak van der Waal's forces, whereas chemisorption involves the formation of relatively strong chemical bonds and requires some activation energy, therefore it may be slow and not always reversible. Above certain temperatures chemisorption dominates and it is at this temperature ALD operates best. Also, chemisorption is the reason that the process is self-controlling and insensitive to pressure and substrate changes because only one atomic or molecular layer can adsorb at the same time.

Equipment for the ALD process

Equipment used in the ALD process may be classified in terms of their working pressure (vacuum, low pressure, atmospheric pressure), method of pulsing the precursors (moving substrate or valve sources) or according to the types of sources. Several system types are discussed.

In a typical moving substrate ALD growth system (Figure 10.11) the substrate, located in the recess part of the susceptor, is continuously rotated and cuts through streams of the gaseous precursors, in this case, trimethylgallium [TMG, $Ga(CH_3)_3$] and arsine (AsH_3). These gaseous precursors are introduced through separate lines and the gases come in contact with the substrate only when it revolves under the inlet tube. This cycle is repeated until the required thickness of GaAs is achieved. The exposure time to each of the gas streams is about 0.3 s.

ALD may be carried out in a vacuum system using an ultra-high vacuum with a movable substrate holder and gaseous valving. In this manner it may be also equipped with an in-situ low-energy electron diffraction (LEED) system for the direct observation of surface atom configurations and other systems such as XPS, UPS, and AES for surface analysis.

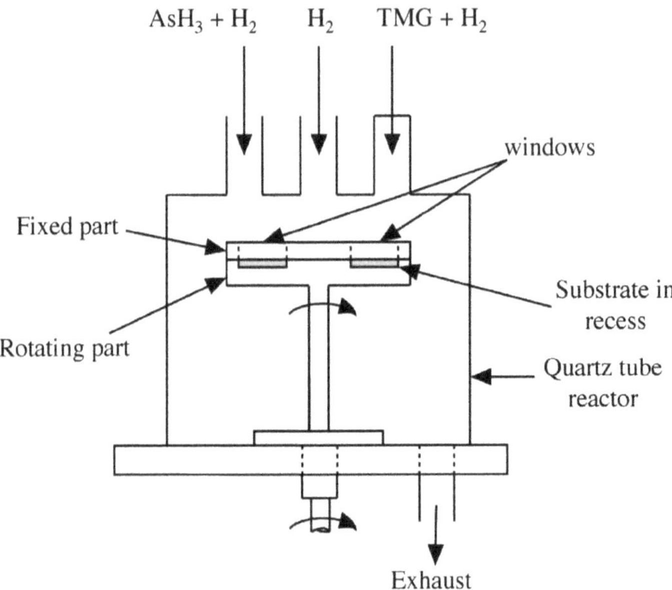

Figure 10.11: A typical moving substrate ALD growth system used to grow GaAs films.

A lateral flow system may also be employed for successful ALE deposition. This uses an inert gas flow for several functions; it transports the reactants, it prevents pump oil from entering the reaction zone, it valves the sources and it purges the deposition site between pulses (Figure 10.12).

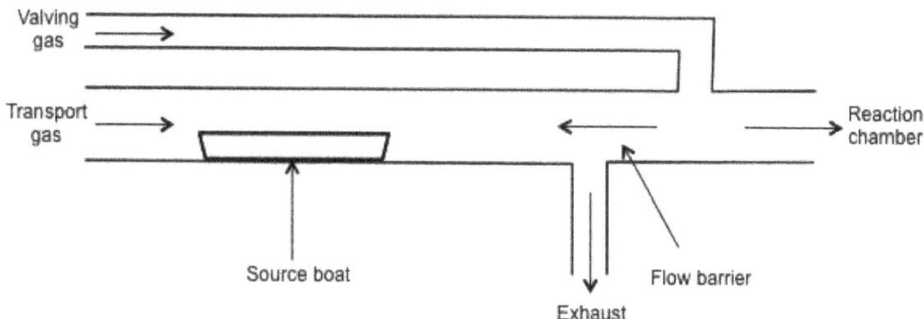

Figure 10.12: Illustration of inert gas valving. Adapted from A. C. Jones and M. L. Hitchman, *Chemical vapour deposition: precursors, processes and applications*. RSC Publishing, London, UK, 2009. Copyright: Royal Society of Chemistry (2009).

Inert gas valving has many advantages as it can be used at ultra high temperatures where mechanical valves may fail and it does not corrode as mechanical valves would in the presence of halides. This method is based on the fact that as the inert gas is flowing through the feeding tube from the source to the reaction chamber, it blocks the flow of the sources. Although in this system the front end of the substrate receives a higher flux density than the down-stream end, a uniform growth rate occurs as long as the saturation layer of the mono-formation predominates. This lateral flow system effectively utilizes the saturation mechanism of a monolayer formation obtained in ALE. Depending on the properties of the precursors used, and on the growth temperature, various growth systems may be used for ALE.

Requirements for ALD growth

Several parameters must be taken into account in order to assure successful ALD growth. These include the physical and chemical properties of the source materials, the pulsing into the reactor, their interaction with the substrate and each other, and the thermodynamics and volatility of the film itself.

Source molecules used in ALD can be either elemental or an inorganic, organic, or organometallic compound. The chemical nature of the precursor is insignificant as long as it possesses the following properties. It must be a gas or must volatilize at a reasonable temperature producing sufficient vapor pressure. The vapor pressure must be high enough to fill the substrate area so that the monolayer chemisorption can occur in a reasonable length of time. Note that prolonged exposure to the substrate can cause the precursor to condense on the surface hindering the growth. Chemical interaction between the two precursors prior to chemisorption on the surface is also undesired. This may be overcome by purging the surface with an inert gas or hydrogen between the pulses. The inert gas not only separates the reactant pulses but also cleans out the reaction area by removing excess molecules. Also, the source molecules should not decompose on the substrate instead of chemisorbing. The decomposition of the precursor leads to uncontrolled growth of the film and this defeats the purpose of ALD as it no longer is self-controlled, layer-by-layer growth and the quality of the film plummets.

In general, temperature remains the most important parameter in the ALD process. There exists a processing window for ideal growth of monolayers. The temperature behavior of the rate of growth in monolayer units per cycle gives a first indication of the limiting mechanisms of an ALD process. If the temperature falls too low, the reactant may condense or the energy of activation required for the surface reaction may not be attained. If the temperature is too high, then the precursor may decompose or the monolayer may evaporate resulting in poor ALD growth. In the appropriate temperature window, full monolayer saturation occurs meaning that all bonding sites are occupied and a growth rate of one lattice unit per cycle is observed. If the saturation density is below one, several factors may contribute to this. These include an oversized reactant molecule, surface reconstruction, or the bond strength of an adsorbed surface atom is higher when the neighboring sites are unoccupied. Then the lower saturation density may be thermodynamically favored. If the saturation density is above one, then the un-decomposed precursor molecules form the monolayer. Generally, ideal growth occurs when the temperature is set where the saturation density is one.

Advantages of ALD

Atomic layer deposition provides an easy way to produce uniform, crystalline, high quality thin films. It has primarily been directed towards epitaxial growth of III-V (13-15) and II-V (12-16) compounds, especially to layered structures such as superlattices and superalloys. This application is due to the greatest advantage of this method, it is controllable to an accuracy of a single atomic layer because of saturated surface reactions. Not only that, but it produces epitaxial layers that are uniform over large areas, even on non-planar surfaces, at temperatures lower than those used in conventional epitaxial growth.

Another advantage to this method that may be most important for future applications, is the versatility associated with the process. It is possible to grow different thin films by choosing suitable starting materials among the thousands of available chemical compounds. Provided that the thermodynamics are favorable, the adjustment of the reaction conditions is a relatively easy task because the process is insensitive to small changes in temperature and pressure due to its relatively large processing window. There are also no limits in principle to the size and shape of the substrates.

One advantage that is resultant from the self-limiting growth mechanism is that the final thickness of the film is dependent only upon the number of deposition cycles and the lattice constant of the material, and can be reproduced and controlled. The thickness is independent of the partial pressures of the precursors and growth temperature. Under ideal conditions, the uniformity and the reproducibility of the films are excellent. ALE also has the potential to grow very abrupt heterostructures and very thin layers and these properties are in demand for some applications such as superlattices and quantum wells.

Chemical vapor deposition

Chemical vapor deposition (CVD) is a deposition process where chemical precursors are transported in the vapor phase to decompose on a heated substrate to form a film. The films may be epitaxial, polycrystalline or amorphous depending on the materials and reactor conditions. CVD has become the major method of film deposition for the semiconductor industry due to its high throughput, high purity, and low cost of operation. CVD is also commonly used in optoelectronics applications, optical coatings, and coatings of wear resistant parts.

CVD has many advantages over physical vapor deposition (PVD) processes such as molecular beam evaporation and sputtering. Firstly, the pressures used in CVD allow coating of three-dimensional structures with large aspect ratios. Since evaporation processes are very directional, PVD processes are typically line of sight depositions that may not give complete coverage due to shadowing from tall structures. Secondly, high precursor flow rates in CVD give deposition rates several times higher than PVD. Also, the CVD reactor is relatively simple and can be scaled to fit several substrates. Ultrahigh vacuum is not needed for CVD and changes or addition of precursors is an easy task. Furthermore, varying evaporation rates make stoichiometry hard to control in physical deposition. While for CVD stoichiometry is more easily controlled by monitoring flow rates of precursors. Other advantages of CVD include growth of high purity films and the ability to fabricate abrupt junctions.

There are, however, some disadvantages of CVD that make PVD more attractive for some applications. High deposition temperatures for some CVD processes (often greater than 600 °C) are often unsuitable for structures already fabricated on substrates. Although with some materials, use of plasma-enhanced CVD or metal-organic precursors may reduce deposition temperatures. Another disadvantage is that CVD precursors are often hazardous or toxic and the by-products of these precursors may also be toxic. Therefore extra steps have to be taken in the handling of the precursors and in the treatment of the reactor exhaust. Also, many precursors for CVD, especially the metal-organics, are relatively expensive. Finally, the CVD process contains a large number of parameters that must be accurately and reproducibly optimized to produce good films.

Kinetics of CVD

A normal CVD process involves complex flow dynamics since gases are flowing into the reactor, reacting, and then by-products are exhausted out of the reactor. The sequence of events during a CVD reaction are shown in Figure 10.13 and as follows:

Step 1. Precursor gases input into the chamber by pressurized gas lines.

Step 2. Mass transport of precursors from the main flow region to the substrate through the boundary layer (Figure 10.13a);

Step 3. Adsorption of precursors on the substrate (normally heated) (Figure 10.13b).

Step 4. Chemical reaction on the surface (Figure 10.13c)

Step 5. Atoms diffuse on the surface to growth sites.

Step 6. Desorption of by-products of the reactions (Figure 10.13d).

Step 7. Mass transport of by-products to the main flow region (Figure 10.13e).

The boundary layer

Gas flow in a CVD reactor is generally laminar, although in some cases heating of the chamber walls will create convection currents. The complete problem of gas flow through the system is too complex to be described here; however, assuming we have laminar flow (often a safe assumption) the gas velocity at the chamber walls will be zero. Between the wall (zero velocity) and the bulk gas velocity there is a boundary layer. The boundary layer thickness increases with lowered gas velocity and the distance from the tube inlet (Figure 10.14). Reactant gases flowing in the bulk must diffuse through the boundary layer to reach the substrate surface. Often, the susceptor is tilted to partially compensate for the increasing boundary-layer thickness and concentration profile.

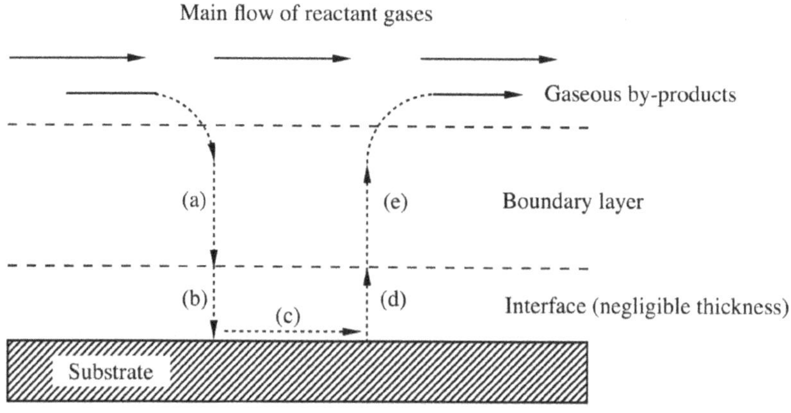

Figure 10.13: Sequence of events during CVD: (a) diffusion of reactants through boundary layer, (b) adsorption of reactants on substrate, (c) chemical reaction takes place, (d) desorption of adsorbed species, and (e) diffusion out of by-products through boundary layer. Adapted from H. O. Pierson, *Handbook of Chemical Vapor Deposition*, Noyes Publications, Park Ridge (1992). Copyright: Noyes Publications (1992).

Rate limiting steps

During CVD the growth rate of the film is limited by either surface reaction kinetics, mass transport (diffusion) of precursors to the substrate, or the feed rate of the precursors.

Surface reaction controls the rate when growth occurs at low temperatures (where the reaction occurs slowly) and also dominates at low pressures (where the boundary layer is thin and reactants easily diffuse to the surface), see Figure 10.15. Since reactants easily diffuse through the boundary layer, the amount of reactant at the surface is independent of reactor pressure. Therefore, it is the reactions and motions of the precursors adsorbed on the surface, which will determine the overall growth rate of the film. A sign of surface reaction limited growth would be dependence of the growth rate on substrate orientation, since the orientation would certainly not affect the thermodynamics or mass transport of the system.

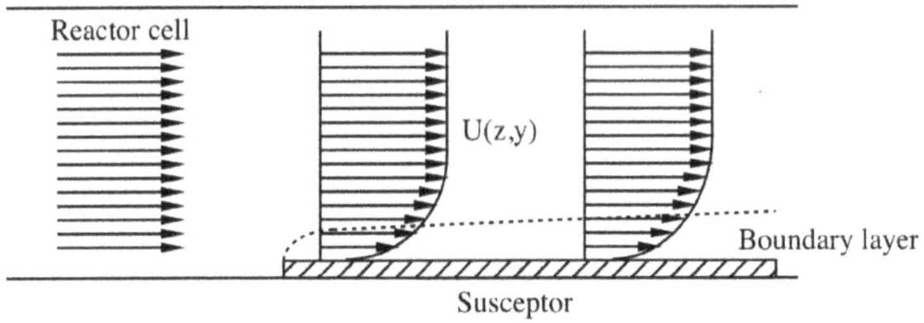

Figure 10.14: Development of boundary layer in a horizontal reactor. Adapted from G. B. Stringfellow, *Organometallic vapor-phase epitaxy: theory and practice*, Academic Press, New York (1994). Copyright: Academic Press (1994).

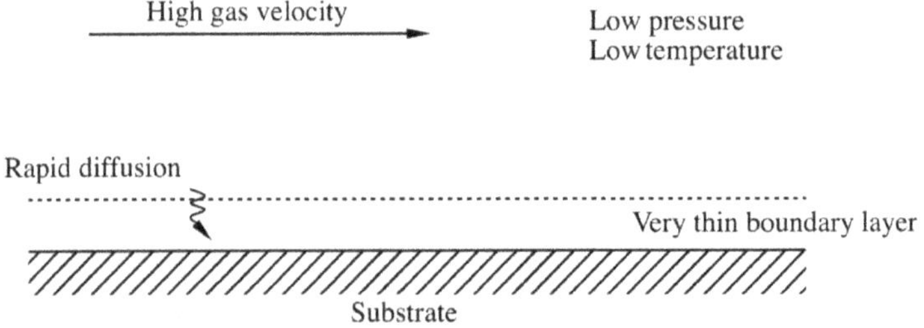

Figure 10.15: Surface reaction limited growth in CVD. Adapted from H. O. Pierson, *Handbook of Chemical Vapor Deposition*, Noyes Publications, Park Ridge (1992). Copyright: Noyes Publications (1992).

A deposition limited by mass transport is controlled by the diffusion of reactants through the boundary layer and diffusion of by-products out of the boundary layer. Mass transport limits reactions when the temperature and pressure are high. These conditions increases the thickness of the boundary layer and make it harder for gases to diffuse through (Figure 10.14). In addition, decomposition of the reactants is typically quicker since the substrate is at a higher temperature. When mass transport limits the growth, either increasing the gas velocity or rotating the substrate during growth will decrease the boundary layer and increase the growth rate.

Feed rate limits the deposition when nearly all the reactant is consumed in the chamber. The feed rate is more important for a hot wall reactor since the heated walls will decompose a large amount of the precursor. Cold wall reactors tend to have higher deposition rates since the reactants are not depleted by reaction at the walls.

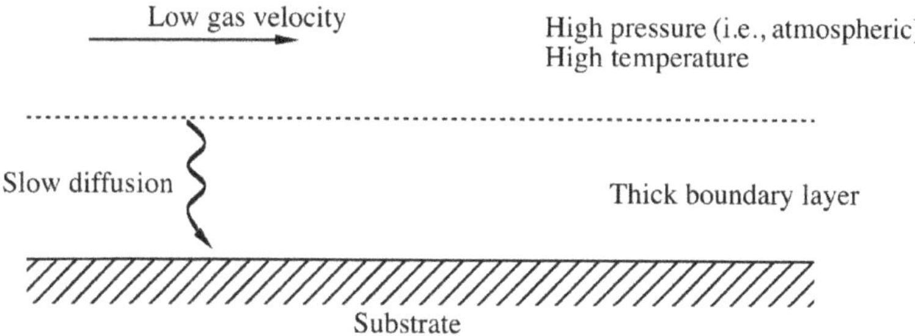

Figure 10.16: Mass transport limited growth in CVD. Adapted from H. O. Pierson, *Handbook of Chemical Vapor Deposition*, Noyes Publications, Park Ridge (1992). Copyright: Noyes Publications (1992).

A plot of growth rate versus temperature, known as an Arrhenius plot, can be used to determine the rate-limiting step of a reaction (Figure 10.17). Mass transport limits reactions at high temperatures such that growth rate increases with partial pressures of reactants, but is constant with temperature. Surface reaction kinetics dominates at low temperatures where the growth rate increases with temperature, but is constant with pressures of reactants. Feed rate limited reactions are independent of temperature, since it is the rate of gas delivery that is limiting the reaction. The Arrhenius plot will show where the transition between the mass transport limited and the surface kinetics limited growth occurs in the temperature regime.

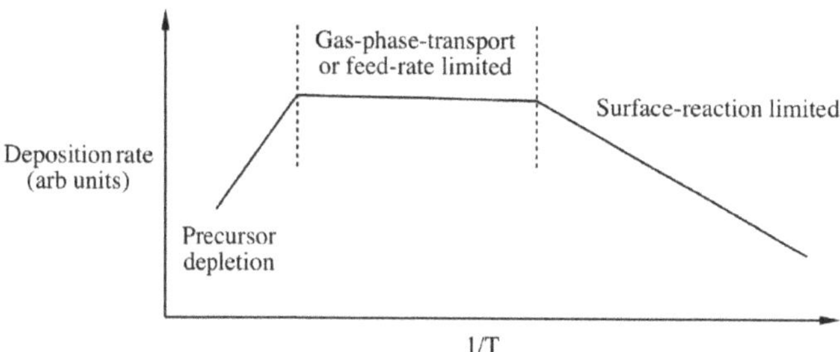

Figure 10.17: Dependence of CVD deposition rate on temperature.

Increases in reactant concentrations will to a point increase the deposition rate. However, at very high reactant concentrations, gas phase nucleation will occur and the growth rate will drop (Figure 10.18). Slow deposition in a CVD reactor can often be attributed to either gas phase nucleation, precursor depletion due to hot walls, thick boundary layer formation, low temperature, or low precursor vapor pressure.

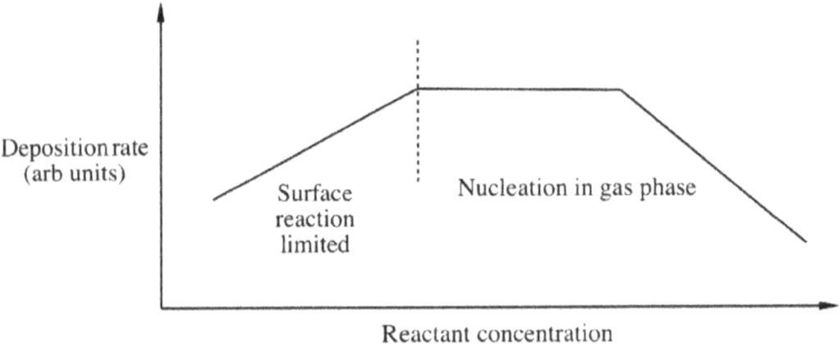

Figure 10.18: Demonstration of deposition rate on reactant concentration for CVD deposition.

CVD systems

Precursor delivery

Flow of reactants into the reactor must be closely monitored to control stoichiometry and growth rate. Precursor delivery is very important since in many cases the flow rate can limit the deposition. For low vapor pressure solids, a carrier gas is passed over or through a bed of the heated solid to transport the vapor into the reactor. Gas flow lines are usually heated to reduce condensation of the vapor in the flow lines. In the case of gas precursors, mass flow meters easily gauge and control the flow rates. Liquid precursors are normally heated in a bubbler to achieve a desired vapor pressure (Figure 10.19).

An inert gas such as hydrogen is bubbled through the liquid and by calculating the vapor pressure of the reactant and monitoring the flow rate of the hydrogen, the flow rate of the precursor is controlled by using,

$$Q_{MO} = Q_{H2} \times \frac{P_{MO}}{P_B - P_{MO}}$$

where Q_{MO} is the flow rate of the metal-organic precursor, Q_{H2} is the flow rate of hydrogen through the bubbler, P_{MO} is the vapor pressure of the metal-organic at the bubbler temperature, and P_B is the pressure of the bubbler.

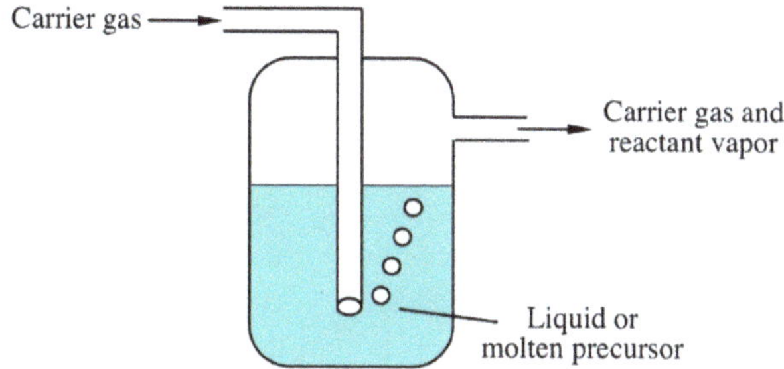

Figure 10.19: Schematic representation of a bubbler for liquid precursors.

Another method of introducing liquid precursors involves flash vaporization where the liquid is passed into a flask heated above the boiling point of the liquid. The gas vapor is then passed through heated lines to the CVD chamber. Often, a carrier gas is added to provide a fixed flow rate into the reactor. This method of precursor introduction is useful when the precursor will decompose if heated over time. A similar technique called spray pyrolysis introduces the precursors in the form of aerosol droplets. The droplets evaporate in the chamber from the heated gas above the substrate or heated chamber walls (Figure 10.20). Then the reaction proceeds as a normal CVD process.

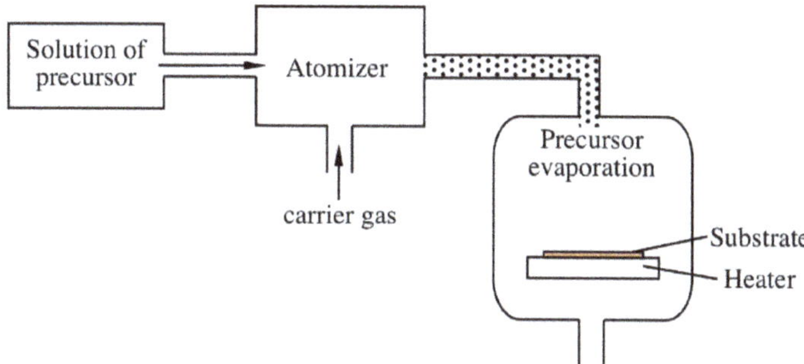

Figure 10.20: Schematic representation of a typical aerosol delivery system for CVD precursors. Adapted from T. T. Kodas and M. J. Hamton-Smith, *The Chemistry of Metal CVD*, VCH, New York (1994). Copyright: VCH (1994).

Thermal CVD reactors

In thermal CVD temperatures as high as 2000 °C may be needed to thermally decompose the precursors. Heating is normally accomplished by use of resistive heating, radio frequency (rf) induction heating, or radiant heating. There are two basic types of reactors for thermal CVD: the hot wall reactor and the cold wall reactor.

A hot wall reactor is an isothermal furnace into which the substrates are placed. Hot wall reactors are typically very large and depositions are done on several substrates at once. Since the whole chamber is heated, precise temperature control can be achieved with correct furnace design. A disadvantage of the hot wall configuration is that deposition occurs on the walls of the chamber as well as on the substrate. As a consequence, hot wall reactors must be frequently cleaned to reduce flaking of particles from the walls, which may contaminate the substrates. Furthermore, reactions in the heated gas and at the walls deplete the reactants and can result in feed rate limited growth. Figure 10.21 shows a typical low-pressure hot wall CVD reactor.

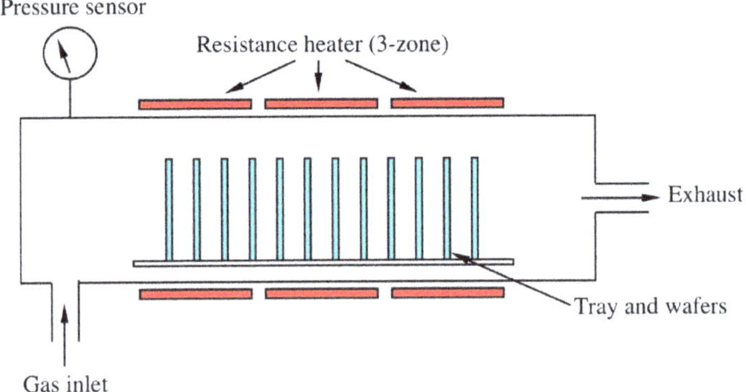

Figure 10.21: Schematic of a typical low-pressure hot wall CVD reactor used in coating silicon substrates. Adapted from H. O. Pierson, *Handbook of Chemical Vapor Deposition*, Noyes Publications, Park Ridge (1992). Copyright: Noyes Publications (1992).

Only the substrate is heated in a cold wall reactor, usually by induction or radiant heating. Since most CVD reactions are endothermic, deposition is preferentially on the area of highest temperature. As a result, deposition is only on the substrate and the cooler reactor walls stay clean. Cold wall CVD has two main advantages over the hot wall configuration. First, particulate contamination is reduced since there are no deposits formed on the walls of the reactor. Second, since decomposition only occurs on the substrate there is no depletion of source gases due to reaction on the walls. However, hot wall reactors tend to have higher throughput since the designs more easily accommodate multiple wafer configurations.

The final issue in design of a thermal CVD reactor is the operating pressure. The pressure of the reactor has a large effect on the rate-limiting step of the deposition. Atmospheric pressure reactors have a large boundary layer (Figure 10.20) and non-uniform diffusion of reactants through the boundary layer often results in non-uniform film compositions across the wafer. Conversely, low-pressure reactors have a nearly non-existent boundary later and reactants easily diffuse to the substrate (Figure 10.19). However, the difficulty in maintaining a uniform temperature profile across the wafer can result in thickness non-uniformities since the deposition rate in low-pressure reactors is strongly temperature dependent. Careful studies of the flow dynamics and temperature profiles of CVD reactors are always carried out in order to achieve uniform material depositions.

Plasma-enhanced CVD

Plasmas are generated for a variety of thin film processes including sputtering, etching, flashing, and plasma-enhanced CVD. Plasma-enhanced CVD (PECVD), sometimes called plasma-assisted (PACVD), has the advantage that plasma activated reactions occur at much lower temperatures compared to those in thermal CVD. For example, the thermal CVD of silicon nitride occurs between 700-900 °C, while the equivalent PECVD process is accomplished between 250-350 °C.

Plasma is a partially ionized gas consisting of electrons and ions. Typical ionization fractions of 10^{-5}-10^{-1} are encountered in process reactors. Plasmas are electrically conductive with the primary charge carriers being the electrons. The light mass of the electron allows it to respond much more quickly to changes in the field than the heavier ions. Most plasma used for PECVD is generated using a radio frequency (rf) electric field. In the high frequency electric field, the light electrons are quickly accelerated by the field, but do not increase the temperature of the plasma because of their low mass. The heavy ions cannot respond to the quick changes in direction and therefore their temperature stays low. Electron energies in the plasma have a Maxwellian distribution in the 0.1-20 eV range. These energies are sufficiently high to excite molecules or break chemical bonds in collisions between electrons and gas molecules. The high-energy electrons inelastically collide with gas molecules resulting in excitation or ionization. The reactive species generated by the collisions do not have the energy barriers to reactions that the parent precursors do. Therefore, the reactive ions are able to form films at temperatures much lower than those required for thermal CVD.

The general reaction sequence for PECVD is shown in Figure 10.22. In addition to the processes that occur in thermal CVD, reactive species resulting from electron dissociation of parent molecules also diffuse to the surface. The reactive species have lower activation energies for chemical reactions and usually have higher sticking coefficients to the substrate. Therefore, the PECVD reaction is dominated by the reactive species on the surface and not any of the parent precursor molecules that also diffuse to the surface.

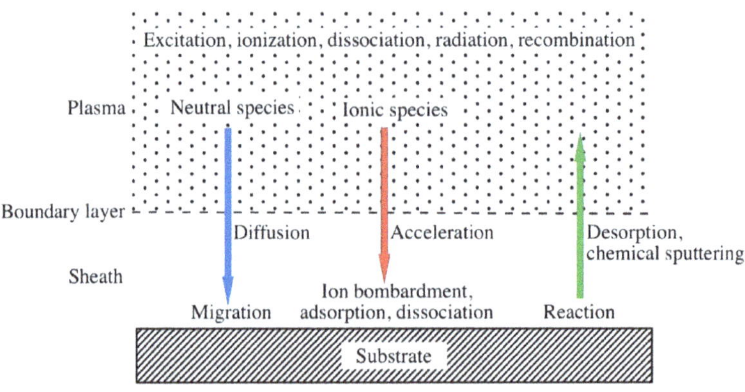

Figure 10.22: Reaction sequence in PECVD. Adapted from M. Konuma, *Film Deposition by Plasma Techniques*, Springer-Verlag, New York (1992). Copyright: Springer-Verlag (1992).

A basic PECVD reactor is shown in Figure 10.23. The wafer chuck acts as the lower electrode and is normally placed at ground potential. Gases are introduced either radially at the edges of the reactor, and pumped out from the center, or introduced from the center and pumped at the edges as shown in Figure 10.23. The magnetic drive allows rotation of the wafers during processing to randomize substrate position. Some newer reactors introduce the gases through holes drilled in the upper electrode. This method of gas introduction gives a more uniform plasma distribution.

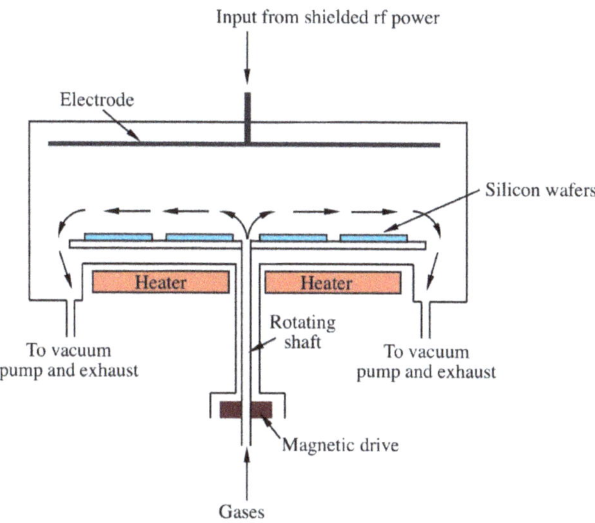

Figure 10.23: Schematic representation of a radial flow PECVD reactor. Adapted from H. O. Pierson, *Handbook of Chemical Vapor Deposition*, Noyes Publications, Park Ridge (1992). Copyright: Noyes Publications (1992).

Plasma CVD has numerous advantages over thermal CVD. Obviously the reduced deposition temperature is a bonus for the semiconductor industry, which must worry about dopant diffusion and metal interconnects melting at the temperatures required for thermal CVD. Also, the low pressures (between 0.1-10 Torr) required for sustaining a plasma result in surface kinetics controlling the reaction and therefore greater film uniformity. A disadvantage of plasma CVD is that it is often difficult to control stoichiometry due to variations in bond strengths of various precursors. For example, PECVD films of silicon nitride tend to be silicon rich because of the relative bond strength of N_2 relative to the Si-H bond. Additionally, some films may be easily damaged by ion bombardment from the plasma.

Photochemical CVD

Photochemical CVD uses the energy of photons to initiate the chemical reactions. Photo-dissociation of the chemical precursor involves the absorption of one or more photons resulting in the breaking of a chemical bond. The most common precursors for photo-assisted deposition are the hydrides, carbonyls, and the alkyls. The dissociation of dimethylzinc [DMZ,

Zn(CH$_3$)$_2$] by a photon creates a zinc radical, and a methyl radical (·CH$_3$) that will react with hydrogen in the reactor to produce methane.

$$Zn(CH_3)_2 + h\nu\,(6.4\,eV) + H_2 \rightarrow Zn + 2\,CH_4$$

Like several metal-organics, dimethylzinc is dissociated by the absorption of only one UV photon; however, some precursors require absorption of more than one photon to completely dissociate. There are two basic configurations for photochemical CVD. The first method uses a laser primarily as a localized heat source. The second method uses high-energy photons to decompose the reactants on or near the growth surface.

In thermal laser CVD, sometimes referred to as laser pyrolysis, the laser is used to heat a substrate that absorbs the laser photons. Laser heating of substrates is a very localized process and deposition occurs selectively on the illuminated portions of the substrates. Except for the method of heating, laser CVD is identical to thermal CVD. The laser CVD method has the potential to be used for direct writing of features with relatively high resolution. The lateral extent of film growth when a laser illuminates the substrate is determined not only by the spot size of the laser, but also by the thermal conductivity of the substrate. A variation of laser pyrolysis uses a laser to heat the gas molecules such that thermal processes fragment them.

A laser can induce photochemical effects if the precursor molecules absorb at the laser wavelength. UV photons have sufficient energy to break the bonds in the precursor chemicals. Laser-assisted CVD (LACVD) uses a laser, usually an excimer laser, to provide the high-energy photons needed to break the bonds in the precursor molecules. Figure 10.24 shows two geometries for LACVD. For the perpendicular illumination the photochemical effects generally occur in the adsorbed adlayer on the substrate. Perpendicular irradiation is often done using a UV lamp instead of a laser so that unwanted substrate heating is not produced by the light source. The parallel illumination configuration has the benefit that reaction by-products are produced further from the growth surface and have less chance of being incorporated into the growing film. The main benefit of LACVD is that nearly no heat is required for deposition of high quality films.

An application of laser photolysis is photonucleation, by which a chemisorbed adlayer of metal precursors is photolyzed by the laser to create a nucleation site for further growth. Photonucleation is useful in promoting growth on substrates that have small sticking coefficients for gas phase metal atoms. By beginning the nucleation process with photonucleation the natural barrier to surface nucleation on the substrate is overcome.

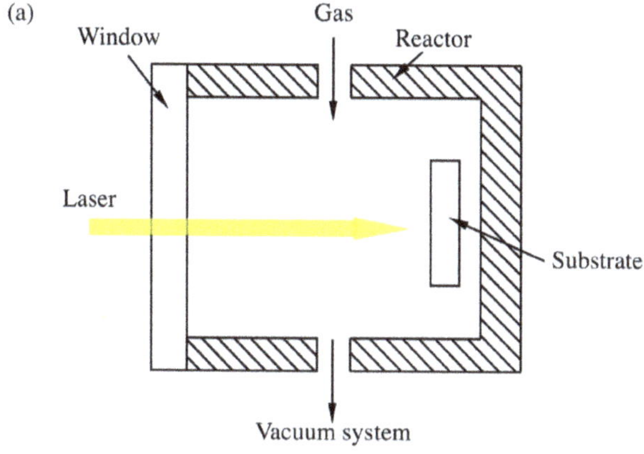

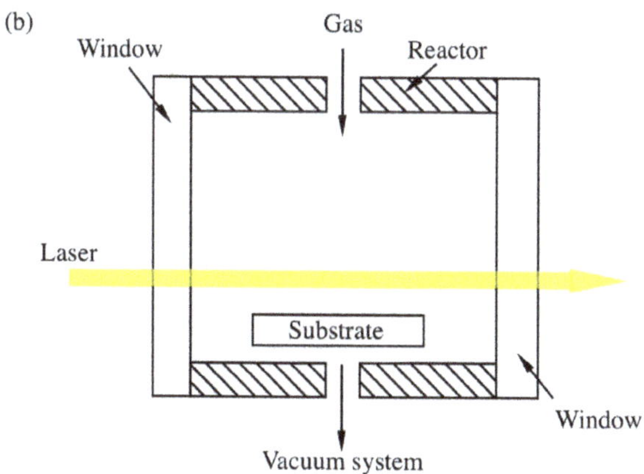

Figure 10.24: Parallel (a) and perpendicular (b) irradiation in laser CVD. Adapted from J. G. Eden, in *Thin Film Processes II*, Eds. J. L. Vossen and W. Kern, Academic Press, New York (1991). Copyright: Academic Press (1991).

Selecting a precursor for chemical vapor deposition

The proven utility of chemical vapor deposition (CVD) in a wide range of electronic materials systems (semiconductors, conductors, and insulators) has driven research efforts to investigate the potential for thin film growth of other materials, including: high temperature superconducting metal oxides, piezoelectric material, etc. Moreover, CVD potentially is well suited for the preparation of thin films on a wide range of substrates, including those of non-planar geometries. CVD offers the advantages of mild process conditions (i.e., low temperatures), control over microstructure and composition, high deposition rates, and possible large scale processing. As with any CVD process, however, the critical factor in the deposition process has been the selection of precursors with suitable transport properties.

Factors in selecting a CVD precursor molecule

The following properties are among those that must be considered when selecting suitable candidates for a CVD precursor:

1. The precursor should be either a liquid or a solid, with sufficient vapor pressure and mass transport at the desired temperature, preferably below 200 °C. Liquids are preferred over solids, due to the difficulty of maintaining a constant flux of source vapors over a non-equilibrium percolation (solid) process. Such non-bubbling processes are a function of surface area, a non-constant variable with respect both to time and particle size. The upper temperature limit is not dictated by chemical factors; rather, it is a limitation imposed by the stability of the mass flow controllers and pneumatic valves utilized in commercial deposition equipment. It must be stressed that while the achievement of an *optimum* vapor pressure for efficient utilization as an industrially practicable source providing high film growth rates (>10 Torr at 25 °C) is a worthy goal, the usable pressure regimes are those in which evaluation can be carried out on compounds which exhibit vapor pressures exceeding 1 Torr at 100 °C.

2. The precursor must be chemically and thermally stable in the region bordered by the evaporation and transport temperatures, even after prolonged use. Early workers were plagued by irreproducible film growth results caused by premature decomposition of source compounds in the bubbler, in transfer lines, and, basically everywhere *except* on the substrate. Such experiences are to be avoided!

3. By its very nature, CVD demands a decomposable precursor. This generally is accomplished thermally; however, the plasma-enhanced growth regime has seen much improvement. In addition, photolytic processes have tremendous potential. Nevertheless, the precursor must be thermally robust *until deposition conditions are employed.*

4. The precursor should be relatively easy to synthesize, ensuring sufficient availability of material for testing and fabrication. It also is important that the synthesis of the compound be reproducible. It should be simple to prepare and purify to a relatively high level of purity. It should be non-toxic and *environmentally friendly* (i.e., as low a toxicity as can be attained, given the fundamental toxicity of particular elements such as mercury, thallium, barium, etc.). It should be routine to reproduce and scale-up the preparation for further developmental studies. It should utilize readily available starting reagents, and proceed by a minimum number of chemical transformations in order to minimize the cost.

5. Due to handling considerations, the source should be oxidatively, hydrolytically, thermally and photochemically stable under normal storage conditions, in addition the precursor should resist oligomerization (in the solid, liquid, or gaseous states). It is worth noting that practitioners of metal organic CVD (MOCVD), especially for III-V materials have of necessity become expert in the handling of very toxic, highly air sensitive materials.

Historically, researchers were limited in their choices of precursors to those that were readily known and commercially available. It must be emphasized that *none* of these previously known compounds had been designed specifically to serve as vapor phase transport molecules for the associated element. Thus, the scope was often limited to what was commercially available; however, as new compounds have now been made with the specific goal of providing ideal CVD precursors the choice to academia and industry has increased.

Liquid phase deposition

The most common forms of silica employed in industry include α-quartz, vitreous silica, silica gel, fumed silica and diatomaceous earth. Synthetic quartz is hydrothermally grown from a seed crystal, with aqueous NaOH and vitreous SiO_2, at 400 °C and 1.7 kbar. Because it is a piezoelectric material, it is used in crystal oscillators, transducers, pickups and filters for frequency control and modulation. Vitreous silica is super cooled liquid silica used in laboratory glassware, protective tubing sheaths and vapor grown films. Silica gel is formed from the reaction of aqueous sodium silicate with acid, after which it is washed and dehydrated. Silica gel is an exceptionally porous material with numerous applications including use as a desiccant, chromatographic support, catalyst substrate and insulator. Pyrogenic or fumed silica is produced by the high temperature hydrolysis, in an oxyhydrogen flame, of $SiCl_4$ (Figure 4.5). Its applications include use as a thickening agent and reinforcing filler in polymers. Diatomaceous earth, the ecto-skeletons of tiny unicellular marine algae called diatoms, is mined from vast deposits in Europe and North America. Its primary use is in filtration. Additional applications include use as an abrasive, insulator, filler and a lightweight aggregate.

Methods of colloidal growth and thin film deposition of amorphous silica have been investigated since 1925. The two most common and well-investigated methods of forming SiO_2 in a sol or as a film or coating are condensation of alkoxysilanes (known as the Stober method) and hydrolysis of metal alkoxides (the Iler or dense silica [DS] process).

Liquid phase deposition

Liquid phase deposition (LPD) is a method for the non-electrochemical production of polycrystalline ceramic films at low temperatures. LPD, along with other aqueous solution methods [chemical bath deposition (CBD), successive ion layer adsorption and reaction (SILAR) and electroless deposition (ED) with catalyst] has developed as a potential substitute for vapor-phase and chemical-precursor systems. Aqueous solution methods are not dependent on vacuum systems or glove boxes, and the use of easily acquired reagents reduces reliance on expensive or sensitive organometallic precursors. Thus, LPD holds potential for reduced production costs and environmental impact. Films may be deposited on substrates that might not be chemically or mechanically stable at higher temperatures. In addition, the use of liquid as a deposition medium allows coating of non-planar substrates, expanding the range of substrates that are capable of being coated. Aqueous deposition techniques have not reached the level of maturation that vapor-phase techniques have in respect to a high level of control over composition, microstructure and growth rates of the resulting films, but their prospect makes them attractive for research.

LPD generally refers to the formation of oxide thin films, the most common being SiO_2, from an aqueous solution of a metal-fluoro complex $[MF_n]^{m-n}$, which is slowly hydrolyzed using water, boric acid $[B(OH)_3]$ or aluminum metal. Addition of water drives precipitation of the oxide. Boric acid and aluminum work as fluoride scavengers, rapidly weakening the fluoro complex and precipitating the oxide. These reactants are added either drop wise or outright, both methods allowing for high control of the hydrolysis reaction and of the solution's supersaturation. Film formation is accomplished from highly acidic solutions, in contrast to the basic or weakly acidic solutions used in chemical bath deposition.

A generic description of the LPD reaction is shown in,

$$H_{(n-m)}MF_n + {}^m/_2\,H_2O \rightleftharpoons MO_{m/2} + n\,HF$$

where m is the charge on the metal cation. If the concentration of water is increased or the concentration of hydrofluoric acid (HF) is decreased, the equilibrium will be shifted toward the oxide. Use of boric acid or aluminum metal will accomplish the latter,

$$H_3BO_4 + 4\,HF \rightleftharpoons BF_4^- + H_3O^+ + 2\,H_2O$$

$$Al + 6\,HF \rightleftharpoons H_3AlF_6 + 1.5\,H_2$$

The most popular of these methods for accomplishing oxide formation has been through the addition of boric acid.

The first patent using liquid phase deposition (LPD) of silicon dioxide via fluorosilicic acid solutions (H_2SiF_6) was granted to the Radio Corporation of America (RCA) in 1950. RCA used LPD as a method for coating anti-reflective films on glass, but the patent promised further applications. Since this initial patent there have been many further patents and papers utilizing this method, in variable forms, to coat substrates, usually silicon, with silicon dioxide. The impetus behind this work is to create an alternative to the growth of insulator coatings by thermal oxidation or chemical vapor deposition (CVD) for planar silicon chip technology.

Thermal oxidation and CVD are performed at elevated temperatures, requiring a higher output of energy and more complicated instrumentation than that of LPD. The most simple and elegant of the LPD methods uses only water to catalyze silica thin film growth on silicon from a solution of fluorosilicic acid supersaturated with silicon dioxide,

$$H_2SiF_6 + 2\,H_2O \rightleftharpoons SiO_{2\downarrow} + 6HF$$

The amount of water reacted with the supersaturated fluorosilicic acid solution controls both the growth rate and incorporation of fluorine into the resulting silica matrix. Both growth rate and fluorine content increase with increased addition of water. Ultimately this "dilution" affects the optical properties of the resulting silica film; an increased amount of fluorine decreases its dielectric constant (and thus its refractive index).

To ensure a uniform film growth with LPD, the preparation of the surface to be coated is of utmost importance. Suitable treatments may involve the formation of surface hydroxides, the pre-deposition or self-assembly of an appropriate seed layer. The most efficient coverage is seen with silicon surfaces functionalized with hydroxy (-OH) groups prior to immersion in the growth solution. This can be achieved through appropriate etching of the silicon surface. It is proposed that the silanol (Si-OH) group acts to seed the growth of the silica film through condensation reactions with the silicic acid formed in the growth solution.

Lee and co-workers and Homma separately propose that intermediate, hydrolyzed species, $SiF_n(OH)_{4-n}$ (n < 4), are formed by the reaction shown in,

$$H_2SiF_6 + (4\text{-}n)\,H_2O \rightleftharpoons SiF_n(OH)_{4\text{-}n} + (6\text{-}n)\,HF$$

According to Lee, these species then react with the substrate surface to form a film. Homma proposes that fluorine-containing siloxanes are subsequently formed, which adsorb onto the surface where condensation and bonding occurs between the oligomers and surface hydroxyl groups. The former mechanism implies a molecular growth mechanism, whereas the latter implies homogeneous nucleation with subsequent deposition.

In concentrated fluorosilicic acid solutions silica can be dissolved to well beyond its solubility, forming fluorosilicon complexes such as $[SiF_6.SiF_4]^{2-}$,

$$5\,H_2SiF_6 + SiO_2 \rightarrow 3\,[SiF_6SiF_4]^{2-} + 2\,H_2O + 6\,H^+$$

The bridged fluorosilicon complex has electron deficient silicon because of the high electronegativity of the bonded fluorines, creating weak Si-F bonds.

These bonds are then prone to nucleophilic attack by water. The fluorine ion (F^-) combines with the proton in this reaction to form hydrofluoric acid (HF). The product of this reaction can then react further with water to yield $[SiF_4(OH)_2]^{2-}$, SiF_4 and HF. The high acidity of the solution then allows protons to react with $[SiF_4(OH)_2]^{2-}$ to form tetrafluorosilicate (SiF_4) and water,

$$[SiF_4(OH)_2]^{2-} + 2H^+ \rightarrow SiF_4 + 2\,H_2O$$

Hydrolysis of the SiF_4 will then yield the hexafluorosilicate anion, protons and silicic acid,

$$3\,SiF_4 + 4\,H_2O \rightarrow 2\,SiF_6^{2-} + Si(OH)_4 + 4\,H^+$$

Silicic acid is adsorbed onto the surface of the substrate that has been introduced into the growth solution. Molecular growth of silica on the substrate surface is initialized in an acid catalyzed dehydration between the silicic acid and the silanol groups on the substrate surface. Si-O-Si bonds are formed, resulting in an initial silica coating of the surface. Following reactions between the initial silica coating and the monosilicic acid in solution result in further silica deposition and growth. Because of the presence of HF in the solution, the surface and growing silica matrix is subject to attack according to the reaction in

$$Si\text{-}OH + HF \rightarrow Si\text{-}F + H_2O$$

This explains the incorporation of a quantity of fluorine into the silica film. Additionally, it reveals that a certain amount of silica etching occurs along with growth. Because of the prevalence of the silicic acid in the solution, however, deposition is predominant.

This proposed mechanism, which is more in depth than those proposed by Lee and Homma, elucidates what is experimentally proven. The deposition rate of the silica increases with addition of H_2O because the nucleophilic attack of the fluorosilicon complex is then augmented, increasing the concentration of silicic acid in the growth solution. The H_2O addition increases the reaction rate and thus the concentration of HF in the growth solution, resulting in greater incorporation of fluorine into the silica matrix because of HF attack of the deposited film. Additionally, Yeh's mechanism supports a molecular growth model, i.e., heterogeneous growth, which represents a consensus of the body of research performed thus far.

In a solution with dissolved ceramic precursors, nucleation and growth will occur either in solution (homogenous nucleation) or on the surfaces of introduced solid phases (heterogeneous nucleation). Successful film formation relies on the promotion of heterogeneous nucleation. Solubility generally depends on the solution pH and the concentration of the species in solution. As the solution crosses over from a solvated state to a state of supersaturation, film formation can occur. It is vital to assure that the state of supersaturation is one that promotes film growth and not homogeneous nucleation and precipitation. This concept is illustrated in Figure 10.25.

Silica can be dissolved in fluorosilicic acid to well above its solubility in water, which is approximately 220 ppm (mg/L). Depending on the concentration of the fluorosilicic acid solution, it can contain up to 20% more silica than is implied by the formula H_2SiF_6. After saturation of the solution with SiO_2, the solvated species is a mixture of fluorosilicates, which reacts as explained earlier. It must be emphasized that addition of water in this reaction is not simply dilution, but is the addition of a reactant, which places the solution in a metastable state (the blue area in Figure 10.23) in preparation for the introduction of a suitable surface to seed the growth of silica.

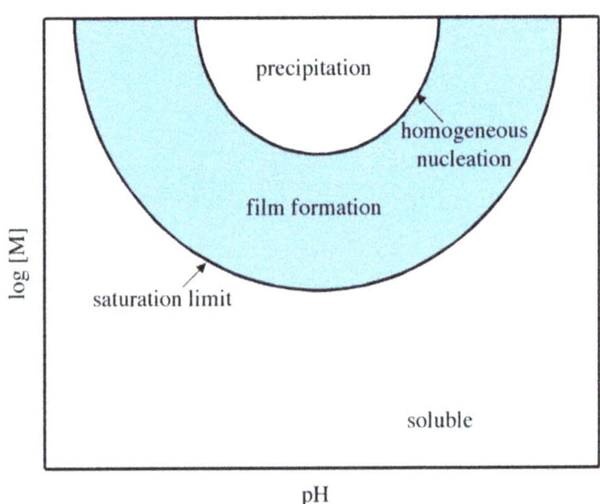

Figure 10.25: Idealized solubility diagram for film forming species in water. Adapted from B. C. Bunker, P. C. Rieke, B. J. Tarasevich, A. A. Campbell, G.

E. Fryxall, G. L. Graff, L. Song, J. Liu, J. W. Virden, and G. L. McVay, Ceramic thin-film formation on functionalized interfaces through biomimetic processing. *Science*, 1994, 264, 48. Copyright: American Association for the Advancement of Science (1994).

Another important factor in solution growth methods is interfacial energy. When a substrate with lower interfacial energy than that of a growing homogeneous nucleus is introduced into a growth solution, heterogeneous growth is favored. Thus, a seeded growth mechanism by definition introduces a substrate of lower interfacial energy into a supersaturated solution, facilitating heterogeneous growth. Lower interfacial energies can be a product of surface modification, as well as a property of the materials' natural state.

Comparing LPD to sol-gel

An alternative method to LPD for forming silica thin films is the sol-gel method. A sol is a colloidal dispersion of particles in a liquid. A gel is a material that contains a continuous solid matrix enclosing a continuous liquid phase. The liquid inhibits the solid from collapsing and the solid impedes release of the liquid. A formal definition of sol-gel processing is the "growth of colloidal particles and their linking together to form a gel." This method describes both the hydrolysis and condensation of silicon alkoxides and the hydrolysis and condensation of aqueous silicates (the DS process).

In the hydrolysis of silicon alkoxides, an alkoxide group is replaced with a hydroxyl group,

$$\equiv Si\text{-}OR + H_2O \rightleftharpoons \equiv Si\text{-}OH + ROH$$

Further condensation reactions between alkoxyl groups or hydroxyl groups produce siloxane bonds:

$$\equiv Si\text{-}OH + RO\text{-}Si\equiv \rightleftharpoons \equiv Si\text{-}O\text{-}Si\equiv + ROH$$

$$\equiv Si\text{-}OH + HO\text{-}Si\equiv \rightleftharpoons \equiv Si\text{-}O\text{-}Si\equiv + H_2O$$

Tetramethoxysilane [$Si(OMe)_4$, TMOS] and tetraethylorthoxysilane [$Si(OEt)_4$, TEOS] are the most commonly used precursors in silica sol-gel processing. The alkoxides are hydrolyzed in their parent alcohols, with a mineral acid or base catalyst, producing silicate gels that can be deposited as coatings. The Stober method, which utilizes this chemistry, relies on homogeneous nucleation to produce monodisperse sols.

The DS method of silica film formation was originally patented as a pigment coating to increase dispersibility of titania particles for use in the paint industry. The DS method is based on the aqueous chemistry of silica and takes advantage of the species present in solution at varying pH. Below pH 7 three-dimensional gel networks are formed. Above pH 7 silica surfaces are quite negatively charged so that particle growth occurs without aggregation:

$$Si(OH)_4 \ (aq) \ \rightarrow \ Si(OH)_3O^- + H^+$$

The isoelectric point of silica is pH 2. Reactions above and below pH 2 are thought to occur through bimolecular nucleophilic condensation mechanisms. Above pH 2 an anionic species attacks a neutral species:

$$SiO^- + \equiv Si\text{-}OH \ \rightarrow \ \equiv Si\text{-}O\text{-}Si \equiv + OH^-$$

While below pH 2 condensation involves a protonated silanol:

$$\equiv SiOH_2^+ + HO\text{-}Si \equiv \ \rightarrow \ \equiv Si\text{-}O\text{-}Si \equiv + H^+$$

The DS process has been utilized extensively in sol-gel coating technology as a growth method for monodisperse and polydisperse sols.

Bibliography

A. C. Adams and C. D. Capio, Edge profiles in the plasma etching of polycrystalline silicon. *J. Electrochem. Soc.*, 1981, **128**, 366.

Advances in Ceramics, Eds. J. A. Mangels and G. L. Messing, American Ceramic Society, Westville, OH, 1984, Vol. 9.

W.M. Alvino, *Plastics For Electronics*, McGraw-Hill, Inc., New York (1995).

A. W. Apblett, L. K. Cheatham, and A. R. Barron, Chemical vapour deposition of aluminium silicate thin films. *J. Mater. Chem.*, 1991, **1**, 143.

D. R. Baigent, N. C. Greenham, J. Gruner, R. N. Marks, R. H. Friends, S. C. Moratti, and A. B. Holmes, Light-emitting diodes fabricated with conjugated polymers - recent progress. *Synth.Met.*, 1994, **67**, 3.

K. G. Baraclough, K. G., in *The Chemistry of the Semiconductor Industry*, Eds. S. J. Moss and A. Ledwith, Blackie and Sons, Glasgow, Scotland (1987).

A. R. Barron, CVD of insulating materials in *CVD of Nonmetals,* Ed. W. Rees, Jr., VCH, 1996.

A. R. Barron, CVD of SiO$_2$ and related materials: An overview. *Adv. Mater. Optics Electron.*, 1996, **6**, 101

A. R. Barron, Cost reduction in the solar industry. *Mater. Today,* 2015, **18**, 2

A. R. Barron and W. S. Rees, Jr., Group 2 compounds as CVD-precursors for electronic materials. *Adv. Mater. Optic. Electron.*, 1993, **2**, 271.

M. Baublitz and A. L. Ruoff, Diffraction studies of the high pressure phases of GaAs and GaP. *J. Appl. Phys.*, 1982, **53**, 6179.

K. J. Bachmann, *The Materials Science of Microelectronics*, VCH (1995).

R. W. Blevins, R. C. Daly, and S. R. Turner, in *Encyclopedia of Polymer Science and Engineering*, Ed. J. I. Krocehwitz, Wiley, New York (1985).

M. J. Bowden, in *Materials for Microlithography: Radiation-Sensitive Polymers*, Ed. L. F. Thompson, C. G. Willson, and J. M. J. Frechet, American Chemical Society Symposium Series No. 266, Washington, D.C. (1984).

A. C. Bonora, *Silicon Wafer Process Technology: Slicing, Etching, Polishing*, Semiconductor Silicon 1977, Electrochem. Soc., Pennington, NJ (1977).

D. C. Bradley, Metal alkoxides as precursors for electronic and ceramic materials. *Chem. Rev.*, 1989, **89**, 1317.

M. Born and E. Wolf, *Principles of Optics 6th Edition*, Pergamon Press, New York (1980).

B. C. Bunker, P. C. Rieke, B. J. Tarasevich, A. A. Campbell, G. E. Fryxall, G. L. Graff, L. Song, J. Liu, J. W. Virden, and G. L. McVay, Ceramic thin-film formation on functionalized interfaces through biomimetic processing. *Science*, 1994, **264**, 48.

J. M. Buriak, L. K. Cheatham, J. J. Graham, R. Gordon, and A. R. Barron, Increased volatility of barium metal organics by the use of nitrogen Lewis bases. *Mat. Res. Soc. Symp. Proc.*, 1991, **204**, 545.

J. H. Burroughes, D. D. C. Bradley, A. R. Brown, R. N. Marks, K. Mackay, R. H. Friend, P. L. Burns, and A. B. Holmes, Light-emitting diodes based on conjugated polymers. *Nature*, 1990, **347**, 539.

P.-H. Chang, C.-T. Huang, and J.-S. Shie, On liquid-phase deposition of silicon dioxide by boric acid addition. *J. Electrochem. Soc.*, 1997, **144**, 1144.

J.-S. Chou and S.-C. Lee, The Initial Growth Mechanism of Silicon Oxide by Liquid-Phase Deposition. *J. Electrochem. Soc.*, 1994, **141**, 3214.

W. M. Cleaver and A. R. Barron, Hybrid organometallic compounds of gallium: $Ga(t\text{-}Bu)_2Me$ and $Ga(t\text{-}Bu)Me_2$. *Chemtronics*, 1989, **4**, 146.

W. M. Cleaver, A. R. Barron, Y. Zhang, and M. Stuke, Hybrid organometallic compounds of gallium: UV excimer laser photochemistry of $Ga(t\text{-}C_4H_9)_n(CH_3)_{3\text{-}n}$ (n = 0, 1, 2, 3). *Appl. Surf. Sci.*, 1992, **54**, 8

L. D. Crossman and J. A. Baker, *Semiconductor Silicon 1977*, Electrochem. Soc., Princeton, New Jersey (1977).

B. H. Cumpston and K. F. Jensen, Photo-oxidation of polymers used in electroluminescent devices. *Synth. Met.,* 1995, **73**, 195.

L. D. Dyer, in *Proceeding of the low-cost solar array wafering workshop 1981*, DoE-JPL-21012-66, Jet Propulsion Lab., Pasadena CA (1982).

J. C. Dyment and G. A. Rozgonyi, Evaluation of a new polish for gallium arsenide using a peroxide-alkaline solution. *J. Electrochem. Soc.*, 1971, **118**, 1346.

J. G. Eden, in *Thin Film Processes II*, Eds. J. L. Vossen and W. Kern, Academic Press, New York (1991).

B. D. Fahlman and A. R. Barron, Substituent effects on the volatility of metal β-diketonates. *Adv. Mater. Optics Electron.*, 2000, **10**, 223.

M. Fleisher, in *Economic Geology, 50th Aniv. Vol.*, The Economic Geology Publishing Company, Lancaster, PA (1955).

H. Gerischer and W. Mindt, The mechanisms of the decomposition of semiconductors by electrochemical oxidation and reduction. *Electrochem. Acta*, 1968, **13**, 1329.

S. K. Ghandhi, *VLSI Fabrication Principles: Silicon and Gallium Arsenide*, 2nd Edition, Wiley-Interscience, NY (1994).

N. C. Greenham, S. C.Maratti, D. D. C. Bradley, R. H. Friend, and A. B. Holmes, Efficient light-emitting diodes based on polymers with high electron affinities. *Nature*, 1993, **365**, 628.

A. Haldar, S. D. Yambem, K.-S. Liao, E. P. Dillon, A. R. Barron, and S. A. Curran, Organic photovoltaics using thin gold film as an alternative anode to indium tin oxide. *Thin Solid Films*, 2011, **519**, 6169.

C. E. Hamilton, D. J. Flood, and A. R. Barron, Thin film CdSe/CuSe photovoltaic on a flexible single walled carbon nanotube substrate. *Phys. Chem. Chem. Phys.*, 2013, **15**, 3930.

C. A. Harper and R. M. Sompson, *Electronic Materials & Processing Handbook*, McGraw Hill, New York, 2nd Edition.

M. D. Healy, P. E. Laibinis, P. D. Stupik and A. R. Barron, The reaction of indium(III)chloride with tris(trimethyl-silyl)phosphine: A novel route to indiumphosphide.. *J. Chem. Soc., Chem. Commun.*, 1989, 359.

M. A. Herman and H. Sitter, *Molecular Beam Epitaxy: Fundamentals and Current Status*, Springer-Verlag (1989).

M. Herold, J. Gmeiner, W. Riess, and M. Schwoerer, Tailoring of the electrical and optical properties of poly (*p*-phenylene vinylene). *Synth. Met.*, 1996, **76**, 109.

T. Homma, T. Katoh, Y. Yamada, and Y. Murao, A selective SiO_2 film-formation technology using liquid-phase deposition for fully planarized multilevel interconnections. *J. Electrochem. Soc.*, 1993, **140**, 2410.

Y. Horikoshi, M. Kawashima, and H. Yamaguchi, Low-temperature growth of GaAs and AlAs-GaAs quantum-well layers by modified molecular beam epitaxy. *Jpn. J. Appl. Phys.*, 1986, **25**, L868.

C.-H. Hsu, J.-R. Wu, Y.-T. Lu, D. J. Flood, A. R. Barron, and L.-C. Chen, Fabrication and characteristics of black silicon for solar cell applications: an overview. *Mat. Sci. Semicon. Proc.*, 2014, **25**, 2.

S. Iida and K. Ito, Selective etching of gallium arsenide crystals in H_2SO_4-H_2O_2-H_2O system. *J. Electrochem. Soc.*, 1971, **118**, 768.

R. K. Iler, *The Chemistry of Silica Solubility, Polymerization, Colloid and Surface Properties, and Biochemistry*, John Wiley & Sons (1979).

S. K. Iya, R. N. Flagella, and F. S. Dipaolo, Heterogeneous decomposition of silane in a fixed bed reactor. *J. Electrochem. Soc.*, 1982, **129**, 1531.

H. R. Jafry, E. A. Whitsitt, and A. R. Barron, Silica coating of vapor grown carbon fibers. *J. Mater. Sci.*, 2007, **42**, 7381.

J. C. Jamieson, Crystal structures at high pressures of metallic modifications of compounds of indium, gallium, and aluminum. *Science*, 1963, **139**, 845.

P. P. Jenkins, A. N. MacInnes, M. Tabib-Azar, and A. R. Barron, Gallium arsenide transistors: realization through a molecular designed insulator. *Science*, 1994, **263**, 1751

A. C. Jones and M. L. Hitchman. *Chemical vapour deposition: precursors, processes and applications*. RSC Publishing, London, UK, 2009.

R. H. Jordan, A. Dodabalapur, L. J. Rothberg, and R. E. Slusher, *Proceeding of SPIE*, 1997, **3002**, 92.

W. Kern, The evolution of silicon wafer cleaning technology. *J. Electrochem. Soc.*, 1990, **137**, 1887.

T. T. Kodas and M. J. Hamton-Smith, *The Chemistry of Metal CVD*, VCH, New York (1994).

M. Konuma, *Film Deposition by Plasma Techniques*, Springer-Verlag, New York (1992).

J. Krauskopf, J. D. Meyer, B. Wiedemann, M. Waldschmidt, K. Bethge, G. Wolf, and W. Schültze, 5th Conference on Semi-insulating III-V Materials, Malmo, Sweden, 1988, Eds. G. Grossman and L. Ledebo, Adam-Hilger, New York (1988).

C. C. Landry, J. Lockwood, and A. R. Barron, Synthesis of chalcopyrite semiconductors and their solid solutions by microwave irradiation. *Chem. Mater.*, 1995, **7**, 699.

Y.-T. Lu and A. R. Barron, Nanopore-type black silicon anti-reflection layers fabricated by a one-step silver-assisted chemical etching. *Phys. Chem. Chem. Phys.*, 2013, **15**, 9862.

Y.-T. Lu and A. R. Barron, Anti-reflection layers fabricated by a one-step copper-assisted chemical etching with inverted pyramidal structures intermediate between texturing and nanopore-type black silicon. *J. Mater. Chem., A*, 2014, **2**, 12043.

Y.-T. Lu and A. R. Barron, In-situ fabrication of a self-aligned selective emitter silicon solar cell using the gold top contacts to facilitate the synthesis of a nanostructured black silicon anti-reflective layer instead of an external metal nanoparticle catalyst. *ACS Appl. Mater. Interfaces.* 2015, **7**, 11802.

Y.-T. Lu and A. R. Barron, Fabrication of anti-reflection coating layers for silicon solar cells by liquid phase deposition. *Main Group Chem.*, 2015, **14**, 279.

A. N. MacInnes, M. B. Power, A. R. Barron, P. P. Jenkins, and A. F. Hepp, Photoluminscence intensity enhancement of GaAs by metal organic chemical vapor deposited GaS from a single source precursor. *Appl. Phys. Lett.*, 1993, **62**, 711.

A. N. MacInnes, M. B. Power, and A. R. Barron, Chemical vapor deposition of cubic gallium sulfide thin films: a new meta-stable phase. *Chem. Mater.*, 1992, **4**, 11.

J. R. McCormic, Conf. Rec. 14th IEEE Photovolt. Specialists Conf., San Diego, CA (1980).

J. R. McCormic, in *Semiconductor Silicon 1981*, Ed. H. R. Huff, Electrochemical Society, Princeton, New Jersey (1981).

J. H. McFee, B. I. Miller, and K. J. Bachmann, Molecular beam epitaxial growth of InP. *J. Electrochem. Soc.*, 1977, **124**, 259.

Y. Mori and N. Watanabe, A new etching solution system, H_3PO_4-H_2O_2-H_2O, for GaAs and its kinetics. *J. Electrochem. Soc.*, 1978, **125**, 1510.

S. J. Moss and A. Ledwith, *The Chemistry of the Semiconductor Industry*, Blackie & Son Limited, Glasgow (1987).

M. Nakase, Recent progress in KrF excimer laser lithography. *IEICE Trans. Electron.*, 1993, **E76-C**, 26.

T. Niesen and M. R. De Guire, Review: deposition of ceramic thin films at low temperatures from aqueous solutions. *J. Electroceramics*, 2001, **6**, 169.

B. L. Oliva and A. R. Barron, Thin films of silica imbedded silicon and germanium quantum dots by solution processing. *Mat. Sci. Semicon.*, 2012, **15**, 713.

B. L. Oliva-Chatelain and A. R. Barron, Experiments towards size and dopant control of germanium quantum dots for solar applications. *AIMS Mater. Sci.*, 2016, **3**, 1.

W. C. O'Mara, Ed. *Handbook of Semiconductor Silicon Technology*, Noyes Pub., New Jersey (1990).

N. Ozawa, Y. Kumazawa, and T. Yao, Effect of seed crystal and composition of solution on the formation of TiO_2 thin film from aqueous solution. *Thin Solid Films*, 2002, **418**, 102.

I. D. Parker and H. H. Kim, Fabrication of polymer light-emitting diodes using doped silicon electrodes. *Appl. Phys. Lett.*, 1994, **64**, 1774.

D. L. Partin, A. G. Milnes, and L. F. Vassamillet, Effect of surface preparation and heat-treatment on hole diffusion lengths in VPE GaAs and $GaAs_{0.6}P_{0.4}$. *J. Electrochem. Soc.*, 1979, **126**, 1581.

M. Pessa, P. Huttunen, and M. A. Herman, Atomic layer epitaxy and characterization of CdTe films grown on CdTe (110) substrates. *J. Appl. Phys.*, 1983, **54**, 6047.

W. G. Pfann, *Zone Melting*, John Wiley & Sons, New York, (1966).

H. O. Pierson, *Handbook of Chemical Vapor Deposition*, Noyes Publications, Park Ridge (1992).

Properties of Gallium Arsenide. Ed. M. R. Brozel and G. E. Stillman. 3rd Ed. Institution of Electrical Engineers, London (1996).

E. Reichmanis, F. M. Houlihan, O. Nalamasu, and T. X. Neenan, in *Polymers for Microelectronics*, Ed. L. F. Thompson, C. G. Willson, and S. Tagawa, American Chemical Society Symposium Series, No. 537, Washington, D.C. (1994).

R. Reif and W. Kern, in *Thin Film Processes II*, Eds. J. L. Vossen and W. Kern, Academic Press, New York (1991).

M. Ritala and M. Leskel, Atomic layer epitaxy - a valuable tool for nanotechnology?. *Nanotechnology*, 1999, **10**, 19.

M. Robbins, J. C. Phillips, and V. G. Lambrecht, Solid solution formation in the systems $CuM^{III}X_2$-$AgM^{III}X_2$ where M^{III}=Al, Ga, In and X_2=S, Se. *J. Phys. Chem. Solids*, 1973, **34**, 1205.

M. Rothschild, A. R. Forte, M. W. Horn, R. R. Kunz, S. C. Palmateer, and J. H. C. Sedlacek, 193-nm lithography. *IEEE J. Selected Topics in Quantum Electronics*, 1995, **1**, 916.

H. Rutledge, B. L. Oliva-Chatelain, S. J. Maguire-Boyle, D. L. Flood, and A. R. Barron, Imbedding germanium quantum dots in silica by means of a modified Stoeber method. *Mat. Sci. Semicon. Proc.*, 2013, **17**, 7.

T. Sakamoto, H. Funabashi, K. Ohta, T. Nakagawa, N. J. Kawai, and T. Kojima, Phase-locked epitaxy using RHEED intensity oscillation. *Jpn. J. Appl. Phys.*, 1984, **23**, L657.

K. Sawada, M. Ishida, T. Nakamura, and N. Ohtake, Metalorganic molecular beam epitaxy of γ-Al_2O_3 films on Si at low growth temperatures. *Appl. Phys. Lett.* **1988**, *52*, 1673.

J. C. Scott, J. Kaufman, P. J. Brock, R. DiPietro, J. Salem, and J. A. Goitia, Degradation and failure of MEH-PPV light-emitting diodes. *J. Appl. Phys.*, 1996, **79**, 2745.

D. W. Shaw, Effects of vapor composition on the growth rates of faceted gallium arsenide hole deposits. *J. Electrochem. Soc.*, 1968, **115**, 777.

F. Snimura, *Semiconductor Silicon Crystal Technology*, Academic Press, New York (1989).

D. Sridevi and K. V. Reddy, Preparation and characterisation of $CuInSe_{2(1-x)}Te_{2x}$ solid solutions. *Mat. Res. Bull.*, 1985, **20**, 929.

G. B. Stringfellow, *Organometallic Vapor-Phase Epitaxy: Theory and Practice*, Academic Press, New York (1994).

W. Stober, A. Fink, and E. Bohn, Controlled growth of monodisperse silica spheres in the micron size range. *J. Colloid Interface Sci.*, 1968, **26**, 62.

T. Suntola and J. Antson, *Method for producing compound thin films*, U.S. Patent 4,058,430 (1977).

A. Szerling, K. Kosiel, M. Płuska, T. Ochalski, and J. Ratajczak, Oval defects in crystals grown by MBE technique: study and methods of elimination. *Electron Technol.*, 2004, **36**, 1.

M. Tabib-Azar, S. Kang, A. N. MacInnes, M. B. Power, P. Jenkins, A. F. Hepp, and A. R. Barron, Electronic passivation of n- and p- type GaAs using chemical vapor deposited GaS. *Appl. Phys. Lett.*, 1993, **63**, 625.

M. A. Tischler and S. M. Bedair, Self-limiting mechanism in the atomic layer epitaxy of GaAs. *Appl. Phys. Lett.*, 1986, **48**, 1681.

W. T. Tsang, Chemical beam epitaxy of $Ga_{0.47}In_{0.53}As/InP$ quantum wells and heterostructure devices. *J. Crystal Growth*, 1987, **81**, 261.

D. R. Turner, On the mechanism of chemically etching germanium and silicon. *J. Electrochem. Soc.*, 1960, **107**, 810.

P. van Zant, *Microchip Fabrication*, 2nd ed., McGraw-Hill Publishing Company, New York (1990).

Y. K. Vohra, S. T. Weir, and A. L. Ruoff, High-pressure phase transitions and equation of state of the III-V compound InAs up to 27 GPa. *Phys. Rev. B*, 1985, **31**, 7344.

M. S. Weaver, D. G. Lidzaey, T. A. Fisher, M. A. Pate, D. O'Brien, A. Bleyer, A. Tajbakhsh, D. D. C. Bradley, M. S. Skolnick, and G. Hill, Recent progress in polymers for electroluminescence: microcavity devices and electron transport polymers. *Thin solid Films,* 1996, **273**, 39.

D. Whitehouse, *Glass of the Roman Empire*, Corning (1988).

E. A. Whitsitt and A. R. Barron, Silica coated single walled carbon nanotubes. *Nano Lett.*, 2003, **3**, 775.

E. A. Whitsitt and A. R. Barron, Silica coated fullerenols: seeded growth of silica spheres under acidic conditions. *Chem. Commun.*, 2003, 1042.

E. A. Whitsitt and A. R. Barron, Effect of surfactant on particle morphology for liquid phase deposition of submicron silica. *J. Colloid Interface Sci.*, 2005, **287**, 318.

R. E. Williams, *Gallium Arsenide Processing Techniques.* Artech House (1984).

C. G. Willson, in *Introduction to Microlithography*, 2nd ed., Ed. L. F. Thompson, C. G. Willson, M. J. Bowden, American Chemical Society, Washington, D.C. (1983).

M. Yan, L. J.Rothberg, F. Papadimitrakopoulos, M. E. Galvin, and T. M. Miller, Defect quenching of conjugated polymer luminescence. *Phys. Rev. Lett.*, 1994, **73**, 744.

S. D. Yambem, A. Haldar, K. S. Liao, E. P. Dillon, A. R. Barron, and S. A. Curran, Optimisation of organic solar cells with thin film Au as anode. *Sol. Energ. Mat. Sol. C.*, 2011, **95**, 2424.

C.-F. Yeh, C.-L. Chen, and G.-H. Lin, The physicochemical properties and growth mechanism of oxide ($SiO_{2-x}F_x$) by liquid phase deposition with H_2O addition only. *J. Electrochem. Soc.*, 1994, **141**, 3177.

W. M. Yin and R. J. Paff, Thermal expansion of AlN, sapphire, and silicon. *J. Appl. Phys.*, 1973, **45**, 1456.

Y. Zhang, W. M. Cleaver, M. Stuke, and A. R. Barron, U.V. excimer laser photo-chemistry of hybrid organometallic compounds of gallium. *Appl. Phys. A.*, 1992, **55**, 261.